THE CHEMICAL SOCIETY
MONOGRAPHS FOR TEACHERS No. 28

Some Aspects of Basic Polymer Science

D. A. BLACKADDER, BSc, MA, DPhil, MIChemE, CEng
Department of Chemical Engineering, University of Cambridge

LONDON: THE CHEMICAL SOCIETY

Monographs for Teachers

This is another publication in the series of Monographs for Teachers which was launched in 1959 by the Royal Institute of Chemistry. The initial aim of the series was to present concise and authoritative accounts of selected well-defined topics in chemistry for those who teach the subject at GCE Advanced level and above. This scope has now been widened to cover accounts of newer areas of chemistry or of interdisciplinary fields that make use of chemistry. Though intended primarily for teachers of chemistry, the monographs will doubtless be of value also to a wider readership, including students in further and higher education.

© *The Chemical Society, 1975*

First published November 1975

Published by The Chemical Society, Burlington House, London W1V 0BN, and distributed by The Chemical Society Publications Sales Office, Blackhorse Road, Letchworth, Herts SG6 1HN

Printed by Adlard & Son Ltd, Bartholomew Press, Dorking

Contents

An inspection of the polymer books on the shelves of a good library will show that more than half of them were published in the last 10 years. This is an indication both of rapidly growing interest in the subject and of the problems facing the author of a new book. However, this monograph is distinctive because of its brevity, and it has been written according to the traditions of the Royal Institute of Chemistry series to which it belongs. The most the reader can expect is a general impression of what polymer science is about, what special principles are involved and, to a lesser extent, what tools are used in polymer research. The treatment presupposes only a very elementary knowledge of chemistry, physics and mathematics. A mathematical relationship is usually the most succinct expression of a scientific principle, yet it remains lifeless without a formal derivation and some discussion of its implications. Derivations have rarely been possible here on grounds of space, but some exceptions to this rule occur in the sections devoted to the kinetics of polymerization reactions. When results are quoted without proof a special attempt is made to stress the context and general usefulness of each relationship.

Polymer science has a distinctive flavour and it is hoped that the reader will become aware of this. Polymers constitute a special class of materials whose behaviour may be partly accounted for in terms of classical chemistry and physics, suitably extended or adapted. In addition, quite new concepts and techniques have been developed to meet the special needs of polymer science in describing the extraordinary architecture and characteristics of long chain molecules. In particular, averaging procedures are of crucial importance when the macroscopic properties of polymer specimens are to be interpreted in molecular terms. Indeed polymer science is permeated by considerations of a statistical or entropic kind.

Naturally occurring polymers are not discussed in this monograph, though they have played an important role in terrestrial affairs before and after the appearance of life on the planet. Some natural polymerizations take place with a speed and precision which the scientist has been unable to match. Inorganic synthetic polymers are also excluded from this discussion where the illustrations will be drawn almost entirely from the realm of synthetic organic materials.

The importance of synthetic polymers in the modern world scarcely needs emphasis, but their present ubiquity is highlighted if the reader will cast his mind back over the day and make a note of every article he has used or noticed which he knows or believes to be of polymeric constitution. Polymers feature in many items of clothing, domestic equipment (especially in the kitchen), wall and floor coverings, soft furnishings, motor vehicles and boats, office and laboratory fittings, toys and buildings, as well as in a multitude of small components in manufactured articles. Even this very general list does not include the less obvious use of polymers in adhesives, paints and coatings. Despite the enormously wide range of products it is important to appreciate that a relatively small number of individual polymers, say a couple of dozen, dominate the market in virtue of their high annual tonnage. The advantages of very large scale production have been realized for these versatile polymers which continue to displace traditional materials where it can be shown that they perform as well or better for lower cost. In the future one would expect to see a progressively more effective use of existing products for all kinds of applications, accompanied by the rather infrequent development of quite new polymers on an initially smaller scale where the higher cost can be justified on the grounds of demonstrably superior properties for a specific application.

The transistor, which became a commercial proposition around 1948, is a familiar example of a device with many practical applications which resulted from some extremely sophisticated theoretical studies. The modern synthetic polymer industry has had a very different kind of history. Until relatively recently polymer technology has been rather poorly supported by polymer science. The true nature of the polymer molecule has been generally recognized for less than 50 years, yet in polymer technology fortunes were being lost, and less commonly made, 100 years ago. However, thanks to an improving relationship between science and technology, the chemical industry enjoyed a period of very well sustained growth after the immediate post-war years and polymers formed the brightest single area. Writing in the middle 1970s it seems that future growth rates may be somewhat smaller than in the buoyant period of 15 years up to 1970, but much better than in the disastrous early 70s. Certainly there will be a steady demand for appropriately qualified manpower at all levels, and some education in polymers and plastics is desirable for everyone now that they have indisputably become part of the scientific and technological facts of life. A plastic is defined as such by common usage rather than by scientific considerations, but it is essentially a mouldable polymer-based material. Those who remember the indifferent plastics produced before and just after the Second World War may have some lingering prejudice against the use of these materials. Education will help people to appreciate that plastic products are attractive not merely on grounds of price, but because they are increasingly the result of good design principles applied to carefully selected materials.

To give the monograph as much cohesion as possible great emphasis has been placed on the polymer molecule.   To the chemist this may well seem entirely natural, but in writing about polymers for scientists in other disciplines it would be perfectly possible to place the emphasis elsewhere, on overall physical properties for example. At any rate the notion of a long chain has been used to bind together the three main sections. The first of these concerns procedures for assembling polymer molecules from monomer units.   This is followed by a discussion of the configurational and conformational properties of polymer molecules, together with the definition and measurement of relative molecular masses.   Finally, aggregates of polymer molecules are classified, and the properties of different kinds of aggregates are interpreted in terms of the underlying structure.   A short summary underlines the attractions of a career in polymer science or technology.

2. Polymerization

General introduction

Before discussing the processes by which giant molecules are now synthesized, it is appropriate to note the role of synthetic studies in the history of the macromolecular hypothesis. This great principle — that polymer molecules are simply very long chains — evolved painfully from a tangle of nineteenth century theories and prejudices. Some of the substances encountered in the early days of colloid chemistry are now known to have been polymers, and the roots of polymer science have much in common with those of colloid science. Polymers and colloids often gave the impression of falling awkwardly and perplexingly outside the framework of mainstream chemistry. Relative molecular mass determinations in particular were distrusted on the grounds that the results for polymers were erratic and unacceptably high. This situation arose because the idea of a macromolecule ran counter to the thinking of many chemists well into the twentieth century. The very real successes of classical synthetic methods led to the common view that molecules of a given substance had a definite structure, were more or less compact, and had relative molecular masses of a few 100 at most. The acceptance of relative molecular masses ranging from 10 000 upwards appeared to require a complete revision of the principles of molecular architecture; a revision which many scientists were unwilling to make.

Amid much controversy the period from 1920 to 1930 saw the gradual acceptance of the macromolecular hypothesis, clearly propounded by Staudinger and comprehensively justified by a variety of ingenious experimental approaches. Towards the end of the decade Carothers made use of the well-tried methods of classical synthesis to build up polymer molecules of known structure and thus secured their recognition as long chain species with a definite repeat unit.

Now that the dust has settled after the years of controversy, it can be seen that the revolution in thought concerning molecules was not so very radical after all: a polymer molecule is distinctive because it is extremely large, *not* because its constituent atoms or groups are linked together in any special way. There is nothing unusual about the chemical valence bonds by which the repeat units are held together in a polymer molecule. Stereochemically and energetically they are perfectly orthodox. Polymerization comes about because the monomer concerned has an appropriate reaction capability which can be channelled towards reaction of a particular type by control of the conditions. The monomers are commonly humble, familiar molecules which happen to have the right qualities for polymeriza-

4

tion one to another. The nature of these coupling reactions will now be discussed in terms of ideal batch processes where the system is perfectly mixed and where composition changes with time. Some industrial processes approximate to this kind of behaviour but continuous operation is also common. For reactors operating continuously, the relationships between residence times, reaction conditions and the relative molecular mass of the product are very interesting and important. In the context of this work, however, it would be unrealistic to assume an appropriate familiarity with the basic theory of continuous operation, hence the more limited kinetic discussion.

Condensation polymerization

There are two main classes of polymerization processes commonly known as condensation and addition polymerization. The first of these results from a sequence of unrelated elementary processes, individually indistinguishable from the corresponding reactions of classical organic chemistry. For example, the esterification reaction may be typified by the elimination of water between benzoic acid and ethanol in the presence of a suitable catalyst.

$$\text{——COOH} + \text{HOC}_2\text{H}_5 \longrightarrow \text{——COOC}_2\text{H}_5 + \text{H}_2\text{O}$$

The ester molecule produced by this reaction is unable to enter into any further esterification reactions because the relevant functional groups have been eliminated. A different situation arises if there is a second functional group on each of the reactant molecules.

$$\text{HOOC——COOH} + \text{HOCH}_2\text{CH}_2\text{OH} \longrightarrow$$

$$\text{HOOC——COOCH}_2\text{CH}_2\text{OH} + \text{H}_2\text{O}$$

When benzene-1,4-dicarboxylic (terephthalic) acid and ethane-1,2-diol (ethylene glycol) react, the ester molecule has functional groups capable of entering into further esterification reactions. The next stage could involve reaction with another molecule of diol or with another molecule of dibasic acid, the product being

$$\text{HOCH}_2\text{CH}_2\text{OOC——COOCH}_2\text{CH}_2\text{OH} \quad \text{or}$$

$$\text{HOOC——COOCH}_2\text{CH}_2\text{OOC——COOH.}$$

These new and more elaborate molecules are still potentially reactive and the process may continue. In this way a molecule of the important fibre-forming polymer poly(ethene terephthalate) is built up by a sequence of unconnected chemical reactions of a familiar type. This is the very essence of condensation polymerization and the status of a polymer molecule is determined by the number of repeat units which it contains. Poly(ethene terephthalate) may be represented by the general formula

$$HO \left[OC - \bigcirc - COOCH_2CH_2O \right]_n H$$

where the brackets enclose the repeat unit. In this type of polymerization the repeat unit in the chain is not identical to the unreacted monomer molecules due to the elimination of a small molecule such as water.

It is obviously a matter of the greatest importance to have some means of relating the extent of chemical reaction, measured in any convenient way, to the distribution of molecular sizes in the system. (The fact that there is always a distribution of sizes should cause no surprise; it is a natural consequence of the random process by which large molecules are assembled.) At first sight the analysis may appear dauntingly difficult, but one approach rests on quite simple probability theory combined with intuition. It is convenient to make use of a simpler reaction than that considered above. A single monomer species may polymerize if it is endowed with both of the necessary functional groups. Thus the ω-hydroxy aliphatic acid $HO(CH_2)_6COOH$ will polymerize to give large molecules represented by

$$H \left[O(CH_2)_6 CO \right]_n OH.$$

To provide an overall characterization of the molecules in the system the Number Average Degree of Polymerization, $\overline{DP}_n$, may be defined as the number of monomer molecules present at the start of the reaction divided by the number of molecules of all sizes present at the stage being considered. Any solvent or catalyst molecules are not included in the summation. Every reacting species present, whether unreacted monomer or high polymer, will have one unreacted COOH group and one unreacted OH group. If the number of COOH groups can be determined, by titration for example, it is easy to establish a connection between the extent of chemical reaction, as measured by the number of unreacted functional groups, and $\overline{DP}_n$.   At time t

$$\overline{DP}_n = \frac{\text{original number of molecules}}{\text{number of molecules of all sizes at time } t} = \frac{\sum_{r=1}^{\infty} rn_r}{\sum_{r=1}^{\infty} n_r} \qquad 1$$

$$= \frac{\text{original number of COOH groups}}{\text{number of COOH groups at time } t} = \frac{N_0}{N_t} \qquad 2$$

Here n_r is the number of r-mers in the system at time t, that is, the number of molecules containing precisely r repeat units. N represents the number of unreacted COOH groups at the time indicated by the subscript.   The summation $\sum_{r=1}^{\infty} rn_r$ includes all units, whether reacted or not, and it is therefore equivalent to N_0, the original number of monomer molecules (or COOH groups) at the start of the reaction.   If p is defined as the fraction of COOH groups reacted at time t, it follows that p increases from 0 to 1 as t increases from 0 to infinity.

$$p = \frac{\text{number of COOH groups reacted at time } t}{\text{original number of COOH groups}}$$

$$= \frac{\text{original number of COOH groups} - \text{number of COOH groups unreacted at time } t}{\text{original number of COOH groups}}$$

$$= \frac{N_0 - N_t}{N_0} = 1 - \frac{N_t}{N_0} \qquad 3$$

Comparing 2 and 3,

$$\overline{DP}_n = \frac{1}{1-p} \qquad 4$$

The next step in the analysis is to relate p to the various individual sizes of molecules present in the system.   This is most easily understood in terms of a hypothetical sampling procedure.   At some given value of p suppose that each reacting molecule in the system, whatever its size, is given a number.   A series of random numbers is then called up and the molecule corresponding to each number is examined as to the number of repeat units which it contains.   The average result is obviously $\overline{DP}_n$, but what is the probability that a molecule selected at random shall be of a particular size, say an r-mer?   An r-mer will contain one unreacted COOH group and $r-1$ linkages involving reacted COOH groups.   The probability of any COOH group being reacted is equal to the fraction of reacted COOH groups in the system as a whole, namely p.   Likewise the probability of any particular COOH group being unreacted is $(1-p)$.   The particular combination of reacted and unreacted groups in an r-mer therefore has probability $p^{r-1}(1-p)$ and is the probability that a molecule selected at random is an r-mer. It must be equivalent to the mole fraction of r-mers in the system.

$$\text{Number fraction of r-mers} = p^{r-1}(1-p) \qquad\qquad 5$$

$$= \text{Mole fraction of r-mers} = \frac{n_r}{\sum\limits_{r=1}^{\infty} n_r} = x_r \qquad\qquad 6$$

$$\left(\sum_{r=1}^{\infty} x_r = (1-p) \sum_{r=1}^{\infty} p^{r-1} = 1\right.$$

as required by the definition of x_r as a fraction$\Big)$.

Comparing equations 1, 5 and 6

$$\overline{DP}_n = \sum_{r=1}^{\infty} x_r r = (1-p) \sum_{r=1}^{\infty} p^{r-1} r = \frac{1}{1-p}$$

The first and last of these identities are already related by equation 4.

The form of the dependence of x_r on r may now be investigated. Inspection of equations 5 and 6 shows that $x_r/x_{r+1} = 1/p$, which is greater than unity for all p. This means mole fraction decreases with increasing size and, in particular, unreacted monomer molecules

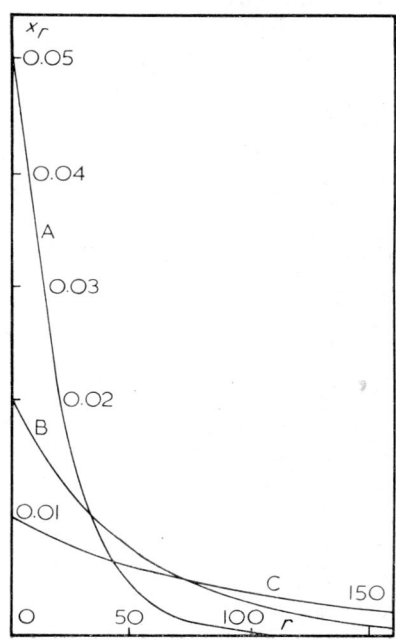

FIG. 1. Condensation polymerization: distribution of relative molecular masses on a number basis. Mole (Number) fraction of r-mers (x_r) for various degrees of polymerization ($\overline{DP}_n = 1/(1-p)$). Curve A, $p=0.95$. Curve B, $p=0.98$. Curve C, $p=0.99$.

are more common than any other species. Very high conversions are evidently necessary to secure a notable proportion of long molecules. *Figure 1* shows how the distribution becomes broader with increasingly high values of p. When $p=0.99$, for example, unreacted monomer molecules account for only 1 per cent of the total on a number basis, yet they are the most common species. In addition there are now enough larger molecules to give the material its polymeric properties.

Equations 4 and 5, involving numbers of molecules of different sizes, are often inadequate as a description of a distribution. They require support from corresponding relationships based on the weight proportions of the various species. Again a hypothetical sampling procedure will help to establish the nature of the new average. Instead of giving each molecule a number it is now necessary to number every unit in the system, whether it is an unreacted monomer molecule or a repeat unit in a polymer chain. As each member of a series of random numbers is selected, the relevant unit is identified and the molecule containing it is examined. This molecule may be simply an unreacted monomer molecule or it may be a polymer molecule of any length. In any event the number of units is noted and an average is calculated from repeated trials. The result is the Weight Average Degree of Polymerization, $\overline{DP}_w$. In this method of averaging the larger molecules are more likely to be selected in the random process because they contain more eligible units; the result is therefore weighted towards the larger molecules and is truly a weight average. The weight fraction of r-mers, w_r, is the probability that a monomer unit, selected at random, is actually part of an r-mer. Using earlier expressions for $\overline{DP}_n$ and x_r,

$$w_r = \frac{rn_r}{\sum\limits_{r=1}^{\infty} rn_r} = \frac{rn_r}{\sum\limits_{r=1}^{\infty} n_r} \cdot \frac{\sum\limits_{r=1}^{\infty} n_r}{\sum\limits_{r=1}^{\infty} rn_r} = \frac{rx_r}{\overline{DP}_n} = rp^{r-1}(1-p)^2 \qquad 7$$

Nothing more is required to calculate the dependence of w_r on r for various values of p. Note that $w_r/w_{r+1}=r/(r+1)p$ and this ratio exceeds unity for all r when p is less than 0.5. (This follows from the fact that $r/(r+1)$ has its minimum value of 0.5 when $r=1$. The differential calculus may *not* be used in this context.) Even on a weight basis the curves will show a continuous fall of w_r with r provided p does not exceed 0.5. For higher values of p a new feature appears, as shown in *Fig. 2*. There is now a maximum value of w_r and this maximum is displaced to higher values of r as p increases. In addition it becomes less pronounced as the distribution broadens.

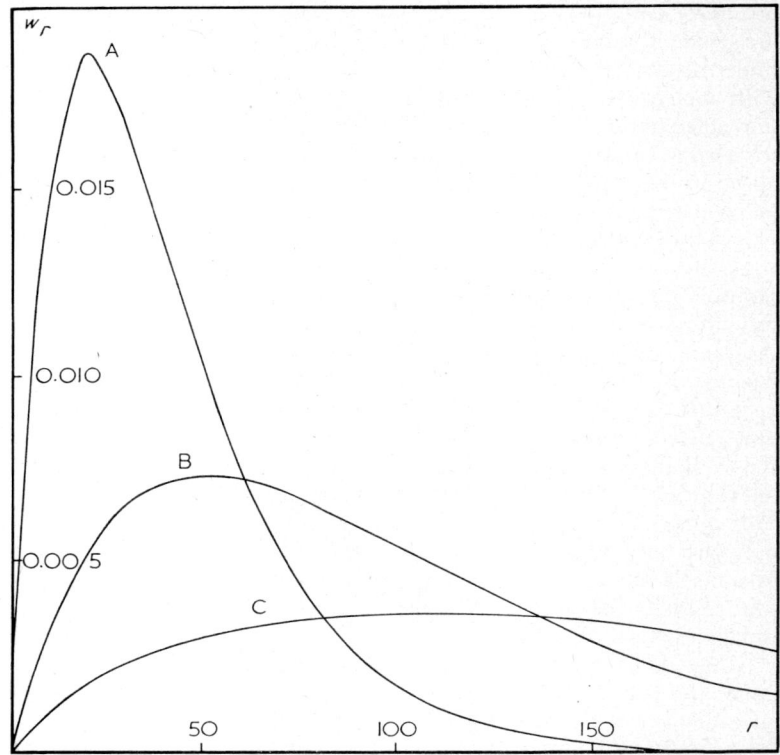

FIG. 2. Condensation polymerization: distribution of relative molecular masses on a weight basis. Weight fraction of r-mers (w_r) for various degrees of polymerization. Curve A, $p=0.95$. Curve B, $p=0.98$. Curve C, $p=0.99$.

The observant reader may have noticed that certain difficulties were glossed over in the above analyses and these now require discussion. In the first place it has been assumed that molecules never react with themselves by cyclization or ring closure. On intuitive grounds it is obvious that cyclization must be extremely improbable for very long molecules so long as there is a good supply of shorter molecules for reactions extending the chain still further. For smaller molecules some inferences may be drawn from the structural chemistry of ring systems. Five-membered rings are known to form with great willingness, while most larger rings form less readily. In addition, detailed experiment has shown that cyclization is not a serious problem provided that the starting monomer is not too small and the system is not too dilute.

It will be recalled that the probability of any particular group being reacted was identified with the total fraction of reacted groups. This assertion carried an important kinetic implication: the reactivity of any particular functional group must not depend on the size of the molecule to which it is attached. If there were such a dependence, say in the sense of functional group reactivity decreasing with increasing chain length, a given value of p would correspond to a size distribution of molecules different from that predicted by equations 5 and 7. It would certainly be more difficult to analyse a system in which reactivity depended on chain length and it is fortunate that in practice the difficulty does not arise. For a non-polymeric esterification reaction it is possible to write

$$\text{Rate} = k[\text{alcohol}]\,[\text{acid}]\,[\text{catalyst}]$$

Flory was able to show that a similar expression held good for a polymerizing system, where each concentration term corresponded to all the available acid or alcohol molecules whatever their size. Reaction between two monomers present in equivalent amounts, for example, is described by

$$-\frac{d[c]}{dt} = k[c]^2\,[c_{\text{cat}}]$$

The catalyst concentration is given by $[c_{\text{cat}}]$, and the reactant concentration, $[c]$, is related to its value at the start of the reaction by $[c] = [c_0](1-p)$. Integration gives

$$\frac{1}{1-p} - 1 = k[c_{\text{cat}}]\,[c_0]t$$

A plot of $\overline{DP}_n$ against t gave a straight line as required.

Finally, among the matters not considered in the general analysis, it might be wondered if condensation polymerization comes to an end in any definite sense. Under given conditions of temperature and catalyst concentration it is obvious that the rate of reaction eventually becomes very slow as the relative molecular mass distribution broadens and the remaining unreacted functional groups take longer and longer to find and react with appropriate partners. In practice the relative molecular mass of the product is often controlled by the exploitation of special chain stopping agents. These molecules have only one functional group and can block the end of a chain to prevent further growth. Acetic acid can block one end of an ester chain:

$$CH_3COOH + H\text{-}[O(CH_2)_6CO]_n\text{-}OH \longrightarrow CH_3CO\text{-}[O(CH_2)_6CO]_n\text{-}OH + H_2O.$$

By adding an appropriate amount of reagent the extent of reaction can be controlled.

Before leaving the subject of linear condensation polymers some mention must be made of nylon, though this is really a general name for a family of polyamides. A nylon results from the reaction of a diamine with a dicarboxylic acid, and the backbone chain contains nitrogen as well as carbon atoms. Hexamethene diamine and hexanedioic (adipic) acid give nylon 66, the numbers corresponding to the carbon atoms in the amine and acid in that order. The polymer is therefore poly(hexamethene adipamide).

$$n\text{H}_2\text{N(CH}_2)_6\text{NH}_2 + n\text{HOOC(CH}_2)_4\text{COOH} \longrightarrow$$

$$\text{H}\text{-}\!\!\left[\text{NH(CH}_2)_6\text{NHCO(CH}_2)_4\text{CO}\right]_n\!\!\text{OH} + n\text{H}_2\text{O}$$

The same amine reacting with decanedioic acid would give nylon 6,10. It is possible to obtain a nylon from a single monomer species provided it has two functional groups, one of each type, though these may be latent. Caprolactam is the starting material for nylon 6, not ε-aminocaproic acid, and the reaction is an example of a stepwise process in which there is no elimination of small molecules.

$$n\text{(CH}_2)_5 \!\!\begin{array}{c}\text{NH}\\ |\\ \text{CO}\end{array} \longrightarrow \text{H}\text{-}\!\!\left[\text{NH(CH}_2)_5\text{CO}\right]_n\!\!\text{OH}$$

Like poly(ethene terephthalate) mentioned earlier, the nylons are very important fibre-forming polymers, much used in clothing, but they are also suitable for the manufacture of rigid components.

The condensation polymers discussed up to this point have been linear chain molecules because each monomer unit had only two functional groups. If some of the monomer molecules have three functional groups a different kind of structure will be produced. Suppose A is a dicarboxylic acid and B is a diol, while C is a trihydroxy alcohol such as propane-1,2,3-triol (glycerol). The incorporation of each molecule of C in a chain produces a branch point as shown in *Fig. 3*, and three dimensional giant molecules extending throughout the polymerizing system will result. At the 'gel point' there is an abrupt increase in the viscosity and a viscous liquid gives way to an elastic gel. At this stage only 10 per cent or so of the monomers may have reacted but this is enough to produce a self-supporting structure with unreacted monomer and higher species embedded in the network. Prolonged reaction times yield a final product which is hard and non-crystalline.

Network polymers with a continuous molecular structure date from the beginning of the century, when Bakelite was first prepared by reacting phenol with methanal. The phenol molecule is effectively trifunctional because it can react at both *ortho* positions and at the *para* position. A water molecule is eliminated on the formation of each methene bridge between benzene rings. Obviously a very tangled and elaborate molecular network will result. The irreversibility of the solidification|process, encouraged by heating, led to the

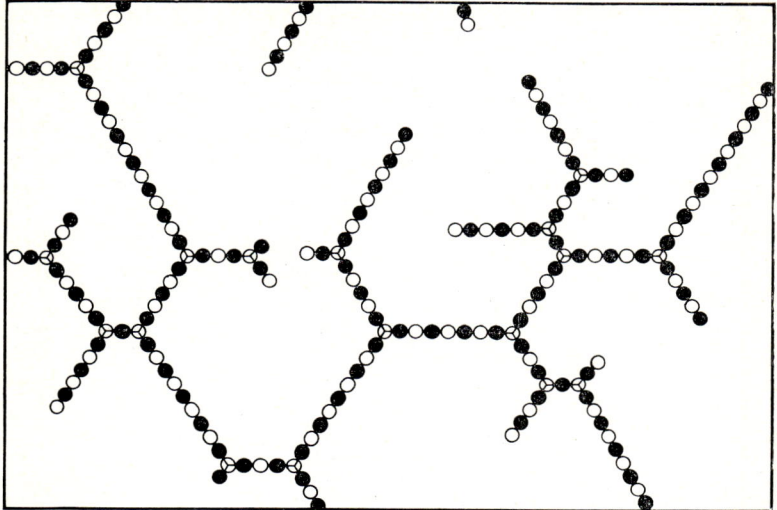

Fig. 3. Two-dimensional representation of a network polymer resulting from the condensation of a dicarboxylic acid ● with a diol ○ and a smaller amount of a trihydroxy alcohol ⊗.

description of these materials as thermosets — to distinguish them from thermoplastics which could be resoftened repeatedly by heating as required.

Addition polymerization

In addition polymerization there is no elimination of a small molecule each time the coupling reaction occurs and the monomer molecules are therefore identical to the repeat units of the polymer chain. Only monomer molecules are eligible for addition to a growing chain and the stages of growth are not separated by the appearance of molecules with reactivities similar to that of the monomer. On the contrary, each polymer molecule reaches its final length rapidly through a sequence of stages, each producing an unstable intermediate. Clearly this has implications for the relative molecular mass distribution. Although long reaction times favour high conversions of monomer to polymer, the length of the polymer chains produced is, to a first approximation, independent of time. From very near the beginning the reaction mixture contains unreacted monomer, some fully grown polymer molecules, and a few chains actually growing. This does not mean that the overall process will be jerky; innumerable over-lapping sequences produce a smooth plot of percentage reaction against time. To appreciate the shape of this overall plot it is necessary to consider the three distinct stages involved in the growth of a given polymer molecule.

Separate initiation is necessary for each sequence of coupling reactions and many types of initiation have been studied and identified. Essentially initiation is a bond-breaking event which produces an active species in the form of a radical or ion. Once initiated, the active species precipitates a sequence of events (chain reaction) leading to the formation of a polymer molecule. The original activity is preserved as each new unit is attached to the growing chain until, sooner or later, some catastrophic event breaks the sequence and that particular polymer molecule reaches the end of its growth history. It must be appreciated that initiation does not normally occur on a once for all basis at the beginning of the complete reaction: it is a continuing process which, in some circumstances, may be assumed to proceed at a constant rate over a significant period of reaction time.

The simplest scheme for addition polymerization appears below where the emphasis is on radicals. Ionic active centres will be discussed in a later section.

$$\text{Initiation process(es)} \qquad \rightarrow P_1$$

$$\text{Propagation steps} \quad \left.\begin{array}{c} P_1 + M_1 \rightarrow P_2 \\ P_2 + M_1 \rightarrow P_3 \\ \cdots\cdots\cdots \\ P_{n-1} + M_1 \rightarrow P_n \end{array}\right\} \text{rate constant } k_p$$

$$\text{Termination} \quad \left\{\begin{array}{l} P_n + P_m \rightarrow M_{n+m} \text{ rate constant } k_{tc} \\ or\ P_n + P_m \rightarrow M_n + M_m \text{ rate constant } k_{td} \end{array}\right.$$

M_1 is a monomer molecule, P_n a growing chain which has already accumulated n units, and M_n, M_m, M_{n+m}, are fully grown (dead) polymer molecules containing the indicated number of repeat units.

It is assumed that k_p is independent of n in all the propagation reactions. This greatly simplifies the kinetic analysis and experiment has shown that the approximation is acceptable. A similar approximation is possible concerning the rate constants k_{tc} and k_{td} for the two forms of the termination reaction. Either two active species give one dead molecule by a process of combination, or their activity is eliminated by the transfer of a hydrogen atom in a disproportionation reaction. Combination is the more common process though both may occur in a given system. The two possibilities may be distinguished if the frequency of 'original chain ends' in the final product can be measured. If combination alone occurs each polymer molecule whether large or small will contain two such ends, whereas disproportionation will result in each product molecule containing only one of the original chain ends. The form of the termination reaction has an obvious bearing on the relative molecular mass of the product. Examples of initiation, propagation and termination stages are shown below for the ethenyl monomer

$CH_2 = CHX$, where X is an atom or group other than hydrogen. It is convenient to let ˜˜˜˜ represent an indeterminate number of linked units.

Initiation. di(benzoyl)peroxide $\longrightarrow$ $2C_6H_5\cdot + 2CO_2$, followed by

$$C_6H_5\cdot + CH_2 =\!\!=\!\!= CHX \longrightarrow C_6H_5CH_2CHX\cdot$$

Propagation. ˜˜˜˜$CH_2CHX\cdot + CH_2 =\!\!=\!\!= CHX \longrightarrow$ ˜˜˜˜$CH_2CHXCH_2CHX\cdot$

Termination by combination.

˜˜˜˜$CH_2CHX\cdot + \cdot CHXCH_2$˜˜˜˜ $\longrightarrow$ ˜˜˜˜$CH_2CHXCHXCH_2$˜˜˜˜

Termination by disproportionation.

˜˜˜˜$CH_2CHX\cdot + \cdot CHXCH_2$˜˜˜˜ $\longrightarrow$

˜˜˜˜$CH =\!\!=\!\!= CHX + CH_2XCH_2$˜˜˜˜

The analysis of the model kinetic scheme on p 14 is straightforward if a steady state may be assumed. This approximation to the truth must not be made too glibly and there are really two related assumptions. The first is, that over a significant part of the complete batch process the total concentration of active species is small and constant. This arises if the rate at which active centres are generated by the initiation process is just balanced by the rate of their removal in the termination process. In kinetic terms:

$$R_i = k_t[P]^2 \qquad\qquad 8*$$

where

$$[P] = \sum_{n=1}^{\infty} [P_n] = (R_i/k_t)^{\frac{1}{2}} \qquad\qquad 9$$

---

* The convention adopted here is that the Rate of Reaction for $aA + bB + \ldots\ldots$ $\overset{k}{\rightarrow}$ products shall be written as $R = k[A]^{\alpha}[B]^{\beta}\ldots\ldots$ where a, b, ... are stoichiometric multipliers and α, β, ... are the orders of reaction with respect to A, B, ... The rate of disappearance (or appearance) of an individual species is related to the Rate of Reaction by

$$-\frac{1}{a}\frac{d[A]}{dt} = k[A]^{\alpha}[B]^{\beta}\ldots etc$$

and similar considerations would apply to the back reaction if any. For the rate-determining initiation reaction $I_2 \overset{k_i}{\rightarrow} 2I\cdot$ (followed by $I\cdot + M_1 \rightarrow P_1$ in a non rate-determining stage),

$$\text{Rate of Reaction} = R = k_i[I_2] = \frac{1}{2}\frac{d[I\cdot]}{dt}$$

The actual rate of production of active centres, the initiation rate, is then given by

$$\frac{d[I\cdot]}{dt} = 2R = R_i$$

The purist must console himself with the thought that the suffix i indicates the rate of production of initiating species, not the rate of the initiation reaction as such!

The other aspect of the steady state assumption is that this state should be attained very rapidly and it has been shown that a few hundredths of a second will normally suffice which is very satisfactory.

It is now possible to sketch the expected form of a curve showing the percentage reaction against time for a typical addition process in a closed system (*Fig. 4,* curve A).   The steady state corresponds to a steady *rate*, and the curve departs from this only when the underlying assumptions fail to hold good as the monomer is consumed.   In practice there is very commonly a spurious delay time or induction period during which the steady rate is gradually established.   This is due to the interference of inhibiting impurities which eliminate active centres prematurely.

A wide range of experimental techniques have been used to study addition polymerization reactions.   The basic requirement is a property which changes to a convenient degree as polymerization occurs so that the rate of consumption of monomer may be evaluated.   Methods commonly used include dilatometry, viscometry and the measurement of refractive index.   In general the initiation pro-

Fig. 4.   Progress of addition polymerization under various conditions.   Curve A.   Ideal, perfectly clean system.   Curve B.   Inhibitor added.

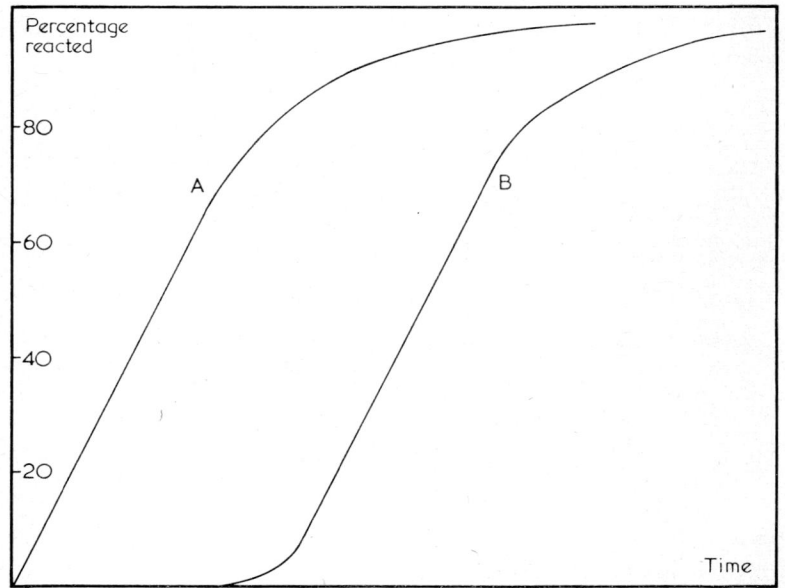

cess makes a negligible contribution to the rate of consumption of monomer which may therefore be identified with the oft-repeated propagation processes

$$-\frac{d[M_1]}{dt} = k_p[M_1][P] = k_p[M_1](R_i/k_t)^{\frac{1}{2}} \qquad 10$$

R_i must be made explicit before the kinetic analysis can advance from the purely symbolic to the quantitatively useful. The detailed form of the expression depends on the circumstances attending the generation of active centres. Thermal initiation may occur if collisional activation suffices to endow some molecules with the appropriate energy but the process is very slow and poorly understood. It sometimes happens that quanta of energy can be directly absorbed by the monomer molecules from incident ultraviolet irradiation. Even if the monomer remains passive it is possible to utilize radiation if a photosensitizer is used. This is a reagent which absorbs the incident quanta and decomposes to give active fragments capable of initiating reaction with the monomer. Although technically inaccurate, the term catalytic initiation is used to describe the most common procedure for initiating addition polymerization. The catalyst is an unstable species which decomposes at an appropriate rate under the reaction conditions to give active fragments which initiate polymerization. The catalyst is consumed by the reaction and, indeed, it is incorporated into the polymer chains. For the initiation process used as an example in the footnote, $R_i = 2k_i f[I_2]$, where $f \leqslant 1$ allows for the possibility that not all the radicals produced by decomposition of the catalyst subsequently attack monomer molecules and initiate chains.

The actual initiation rate under prescribed conditions may be measured using a deliberately added inhibitor, if the stoichiometry of the reaction is properly understood. An inhibitor reacts preferentially with active centres virtually as soon as they are formed and eliminates them from the system. For a fixed amount of inhibitor no visible reaction occurs for a while and then, when the inhibitor is largely consumed, the reaction against time plot starts to assume its normal shape but displaced along the time axis. Extrapolation of the steady rate will give the time during which all active centres were suppressed on formation (curve B, *Fig. 4*). The number of centres suppressed is calculable from the number of inhibitor molecules, hence R_i in appropriate units.

Another quantity of interest is the kinetic chain length, v, defined as the number of monomer units caused to react by the generation of one active centre in the system. A convenient dynamic definition of this pure number is the ratio of the rate of monomer consumption to the rate of initiation. Since the rate of monomer consumption is essentially the rate of propagation as explained above,

$$\nu = \frac{k_{\mathrm{p}}[M_1][P]}{R_i} = \frac{k_{\mathrm{p}}[M_1]}{(R_i k_t)^{\frac{1}{2}}} \qquad 11$$

At first sight it might appear that the kinetic chain length must be simply related to the number average degree of polymerization of the polymer molecules being produced (monomer excluded from the average). This is sometimes the case but not always. During polymerization it often happens that the growth of a particular chain is arrested by an event which, unlike the normal termination, does not eliminate an active species from the system as a whole. Instead, the activity is transferred elsewhere and another distinct chain may grow. Transfer may involve unreacted monomer or fully grown polymer, as well as solvent or chain transfer agent if present. An atom is abstracted by the original radical which dies in the process and a new radical is created at the site of abstraction. The following examples typify transfer to monomer, polymer and solvent respectively.

~~~~~$CH_2CHX\cdot + CH_2{=}CHX \longrightarrow$ ~~~~~$CH_2CH_2X + CH_2{=}CX\cdot$

~~~~~$CH_2CHX\cdot +$ ~~~~~ $\longrightarrow$ ~~~~~$CH_2CH_2X +$ ~~~~~$\cdot$~~~~~

(leading to side branch formation)

~~~~~$CH_2CHX\cdot + CCl_4 \longrightarrow$ ~~~~~$CH_2CHXCl + CCl_3\cdot$

Transfer does not affect the overall rate of monomer consumption but it has a profound effect on the relative molecular mass distribution of the product.

Whatever the transfer reactions in a system, $-d[M_1]/dt$ and R_i are directly measurable, and ν is calculable since $[M_1]$ is known. Inspection of equations 10 and 11 shows that k_{p} and k_t occur only in combination, and special techniques are necessary to obtain separate values of these constants. They have been measured for different chain lengths and the assumption that they are constant is apparently reasonable.

Many of the most important commercial products are addition polymers such as poly(ethene), poly(propene), [poly(phenylethene)], poly(1-chloroethene) and poly[1-(methoxycarbonyl)-1-methylethene].* Their properties, discussed later, cover an enormously

---

* These and other polymers are still largely known by names no longer approved, *viz.* polyethylene, polypropylene, polystyrene, polyvinylchloride or pvc, and polymethyl methacrylate. Names which are strictly proprietary, such as polythene and perspex, are also commonly used.

wide range and composite products are available whose properties lie between those associated with two or more pure polymers. If two different polymers are blended together, the resulting physical mixture will obviously possess properties which relate to those of the pure materials. Adequate blending is not easily secured, however, and a more intimate chemical mixing is usually preferred.

The starting point for the important process of copolymerization is a mixture of two (or more) distinct monomers which are polymerized together. When certain basic requirements are satisfied, each molecule of product will contain some units of each type; contrasting sharply with the purely physical mixture obtained from pre-formed individual polymers. The relationship between the composition of copolymer being formed and the instantaneous composition of the reaction mixture is a problem of some subtlety. During the propagation stage of the reaction the ends of some of the growing chains will bear units of type A and some will be of type B. In principle each type of chain end can add an A or a B monomer molecule when next it reacts. Four reaction rate constants are therefore necessary for a full description, even assuming that chain length is unimportant. The rate of any reaction depends on concentration as well as on the rate constants, hence, to some extent, the course of a copolymerization can be controlled by means of the monomer concentrations. Only in special circumstances where the four rate constants have a unique relationship will the polymer constitution remain equal to that of the monomer mixture as reaction proceeds. Normally the polymer being formed at any instant will not have the same composition as the parent mixture of reacting monomers and both compositions will change continuously. Styrene [$CH_2=CHC_6H_5$] and methyl methacrylate

$$\left[\begin{array}{c} CH_3 \\ | \\ C=CH_2 \\ | \\ COOCH_3 \end{array} \right]$$

are unusual in that they form a copolymer which is almost truly random and whose composition therefore remains closely similar to that of the reaction mixture as reaction proceeds. On the other hand styrene and *cis*-butenedioic (maleic) anhydride

$$\left[\begin{array}{c} CH=CH \\ | \quad\quad | \\ CO \quad CO \\ \diagdown \; O \; \diagup \end{array} \right]$$

have a marked preference for coupling alternately on electronic grounds, and they tend to do so unless styrene is present in very large excess. Some copolymers are of great commercial importance such as the styrene–butadiene rubber used in tyres for motor vehicles, and ABS (acrylonitrile–butadiene–styrene) used for many strong rigid mouldings. The latter contains units of three distinct types.

Block and graft copolymers are decidedly non-random and exploit the use of chemical linkages to bind together rather dissimilar chains. Networks also figure in the exploitation of addition polymerization and two examples follow. Natural rubber is based on isoprene (methylbuta-1,3-diene)

$$\left[CH_2{=}C{-}CH{=}CH_2 \atop \qquad | \atop \qquad CH_3 \right]$$

which undergoes 1,4 addition in the *cis* form to give the polymer chains. The resulting product molecules are unsaturated and oxidizable, the repeat unit being

$$\left[CH_2{-}CH{=}CH{-}CH_2 \atop \quad\;\; | \atop \quad\;\; CH_3 \right]$$

and they are also mechanically unacceptable for heavy wear. Goodyear made the important discovery that sulphur could 'Vulcanise' rubber by forming cross links between the chains, thus forming a strong polymer network, and destroying the double bonds at the same time.

Reinforcement of polymers by glass fibre greatly extends the range of applications of the polymers concerned, and the use of these reinforced plastics involves network polymerization in an interesting and ingenious way. An unsaturated polyester, such as that formed from *cis*-butenedioic acid and ethane-1,2-diol,

$$HO{-}\left[OCCH{=}CHCOOCH_2CH_2O\right]_n H,$$

may be dissolved in a liquid such as styrene monomer to form a syrup. It is then possible to induce cross link formation between the chains by activating the styrene monomer. This second stage of polymerization may be initiated by an organic peroxide catalyst in the form of a non-explosive paste or dispersion easily mixed with the syrup. If glass fibre is quickly impregnated with the mixture and held in a suitable mould excellent components of high strength may be produced. The polymerization is called curing and has three distinct stages.   The gel time is the period required to produce a soft gel, and the hardening time gives the product sufficient hardness and coherence for it to be removed from the mould.   Finally, the so-called maturing time is much less definite but corresponds to virtually complete reaction where the properties of a hard resin are fully realized.   It may be a matter of hours, days or even weeks.   The process does not involve the use of elevated pressures during mould-ing and it is even possible to obviate the need for any external heating to speed the curing.   This is achieved using an accelerator which works with the catalyst to ensure that most of the reaction occurs quickly with a useful evolution of heat *in situ*.   When properly controlled the process is remarkably clean and free from side-products.   The polyester syrup is not stable indefinitely but it might have a shelf life of months or even years before transforming into a gel spontaneously.   The whole procedure is a most effective combination of condensation polymerization (to produce the unsaturated polyester) and addition polymerization (to insert the cross links).

Ionic polymerization

Ionic polymerization is essentially an addition process but it is important enough, and distinctive enough, to justify a separate section.   A number of rather diverse techniques are covered by this heading, and it is convenient to distinguish between them broadly on the basis of the number of phases involved.

Homogeneous ionic polymerization is brought about in solution by means of appropriate initiators.   These species, not necessarily ionic, have the power to couple repeatedly with monomer molecules to produce ever longer chains which retain the ionic characteristics of the original.   The ions are highly reactive but not very selective, and small amounts of impurity will deactivate them.   Systems of very high purity are therefore required, in contrast to condensation polymerization where molecules with irrelevant functional groups can be tolerated.   The batchwise production of polymer by a homogeneous ionic process has three stages: preparation of the initiator system, polymerization, and termination.   The polymeriza-tion of styrene provides an interesting example of a 'living' polymer where termination must be enforced.

A suitable initiator solution may be prepared by reacting a solution of naphthalene in an inert solvent, such as dioxane or tetrahydrofuran, with sodium metal.   The resulting green solution is then mixed rapidly

$$C_{10}H_8 + Na \longrightarrow Na^+C_{10}H_8^-$$

with a solution of styrene monomer in the same solvent. Each initiator species reacts with a styrene molecule to form a radical ion which rapidly dimerizes. The dimer may be regarded as the starting point for the

$$Na^+C_{10}H_8^- + C_6H_5CH{=}CH_2 \longrightarrow Na^+CH^-{-}CH_2\cdot + C_{10}H_8$$
$$\overset{|}{C_6H_5}$$

$$2Na^+CH^-{-}CH_2\cdot \longrightarrow Na^+CH^-CH_2CH_2CH^-Na^+$$
$$\overset{|}{C_6H_5} \qquad\qquad \overset{|}{C_6H_5} \quad \overset{|}{C_6H_5}$$

main propagation process. Growth can occur from both ends and additional monomer units join the chain by slipping in between the charged species with subsequent charge migration. The oppositely charged ions often stay close together because of the low permittivity of the medium. A typical intermediate is represented by

$$Na^+CH^-CH_2\!\left[\!\overset{|}{CHCH_2}\!\right]_m\!\left[\!CH_2\overset{|}{CH}\!\right]_n\!CH_2CH^-Na^+$$
$$\overset{|}{C_6H_5}\quad \overset{|}{C_6H_5}\qquad \overset{|}{C_6H_5}\qquad \overset{|}{C_6H_5}$$

These ionic species are orange-red which has the practical advantage that accidental contamination of the system discharges the colour and is easily recognized. There is no intrinsic termination reaction, assuming the system is sufficiently pure, and deliberate action is necessary to kill off the so-called living ends of the polymer chains. If reaction were allowed to proceed until all of the monomer was used up then, in principle, the chain ends would persist in a state of suspended animation. The addition of fresh monomer at some later time would lead to further growth on the original chains. Indeed it is even possible to introduce a different monomer and so create block copolymer molecules of the type AAAA . . . . AAAABBBBB . . . . . BBBB. For the styrene polymerization considered here, termination is normally achieved by adding water or methanol thus

$$\cdots{-}CH^-Na^+ + H_2O \longrightarrow \cdots{-}CH_2 + NaOH$$
$$\overset{|}{C_6H_5} \qquad\qquad \overset{|}{C_6H_5}$$

This type of polymerization, occurring in three stages which are individually controllable to a considerable degree, has important applications. A product with a very narrow relative molecular mass distribution may be obtained; commonly described as a sharp fraction even though it has resulted directly from polymerization rather than from laborious fractionation of a highly polydisperse specimen. Success depends on rapid mixing of the solutions. In the first place the initiator and monomer must encounter one another so rapidly that virtually all the chains start to grow at the same time and at the same rate. Rapid termination, achieved by introducing a quench solution, is carried out before all the monomer is used up and chain growth is stopped everywhere. For the polymer molecules alone

$$\overline{DP}_n = \frac{\text{number of monomer molecules reacted}}{\frac{1}{2} \text{ number of initiator complexes}}$$

The spread of relative molecular masses is expressed by the ratio $\overline{DP}_w/\overline{DP}_n$, and values as small as 1.04 have been recorded. These should be compared with ratios of 5, 10 or even more resulting from less controllable addition polymerization procedures. Ideally the sharp fraction will have a Poisson distribution of relative molecular masses. Anionic polymerizations of this type give the best fractions when carried out on a laboratory scale, but for many investigations into the effect of relative molecular mass on polymer properties relatively small quantities are quite adequate.

One of the most important groups of ionic polymerization procedures is that based on Ziegler–Natta catalysts. Ziegler discovered a method of polymerizing ethene which did not require the high temperatures and pressures necessary for the free radical addition process. His catalysts were complicated organometallic substances of very high activity. When Natta used similar catalysts for the polymerization of α-alkenes, such as polypropene, an astonishing property came to light. The catalysts were stereodirecting. Instead of the propene molecules joining up randomly as in other processes for this monomer, polymer molecules of a high degree of stereochemical purity were produced. The terms atactic, isotactic and syndiotactic were coined by Natta to denote random, constant and alternating configurations along the chain. These possibilities are shown schematically in *Fig. 5* and derive directly from the fact that one carbon atom in each repeat unit is a pseudo asymmetric centre.

Typically there are three phases present in a Ziegler–Natta system. For example, if propene gas is maintained at a fixed pressure it will ensure that there is a constant concentration of propene monomer in an inert liquid diluent such as heptane. The catalyst may be a soluble alkyl such $Al(C_2H_5)_3$ adsorbed on suspended particles of solid $TiCl_3$. Reaction occurs at certain points on the surface of these particles and the polymer molecule grows as a result of monomer units joining the chain at the point where it is attached to the catalyst particle. A considerable variation in activity exists over the catalyst, and a very broad distribution of relative molecular masses is commonly obtained. The precise nature of the catalyst action is still under discussion and investigation.

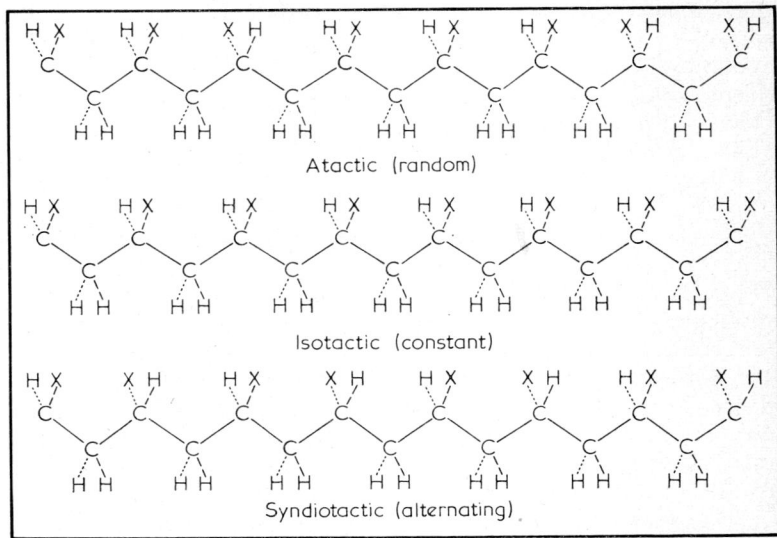

FIG. 5. Schematic representation of stereochemical possibilities for a polymer based on $CH_2=CHX$, where X is any atom or group other than hydrogen. ------- bonds behind plane of paper. — — — bonds in front of paper.

Any catalyst with a solid surface can become fouled, so soluble Ziegler–Natta catalysts have been developed which do not suffer from this disadvantage. Although of high activity they are not necessarily stereodirecting. These catalysts are very useful, notably for the preparation of random copolymers of ethene and propene.

3. The Polymer Molecule

Causes of heterogeneity in polymers

From the discussion of polymerization processes it will be clear that a sample of a synthetic polymer is never pure in the homomolecular sense. Even a sharp fraction, however obtained, will have a spread of relative molecular masses, while in the chaotic conditions of addition polymerization at high temperatures and pressures highly branched molecules are likely to form as a result of extensive chain transfer. The remarkable Ziegler–Natta catalysts do not completely exclude the heterogeneity associated with irregular stereochemical configurations along a given polymer chain.

Apart from these major factors preventing molecular purity in a polymer sample there are others which it is convenient to summarize here. It is not wholly true that a monomer of the type $CH_2=CHX$ polymerizes in a head-to-tail fashion throughout. The fact that this form of addition is preferred does not mean that the other possibility is excluded. A higher activation energy for head-to-head coupling may be partly offset by a more favourable entropy term. When this form of coupling does occur, the chain concerned acquires a point of non-uniformity.

$$\ldots\ldots CH_2CHXCH_2CHXCH_2CHXCHXCH_2CHXCH_2CHXCH_2\ldots\ldots$$

The end of a polymer chain is inevitably different from the repeat units. At the very least there will be a persisting double bond or an extra atom providing saturation, but more probably there will be foreign species present as a result of the participation of such species in the initiation and termination reactions.

Copolymers represent a special case of molecular heterogeneity, except for the perfectly regular alternation of species sometimes observed.

Relative molecular masses

The molecular heterogeneity of a typical polymer sample requires a careful reconsideration of the familiar term 'Relative Molecular Mass'. Neglecting end-groups, the relative molecular mass of an individual polymer molecule is unambiguous, being the product of the number of repeat units and the relative mass of one such unit. However, with rare exceptions, it is not possible to scrutinize individual molecules and some kind of average relative molecular mass must be used to describe a given sample of polymer. There is really no experimental difference between measuring relative molecular masses and degrees of polymerization, since these quantities are related through the relative mass of the repeat unit.   In particular $\bar{M}_n = m\overline{DP}_n$ and $\bar{M}_w = m\overline{DP}_w$, where $\bar{M}_n$ and $\bar{M}_w$ are the number and weight average relative molecular masses and m is the relative mass of the repeat unit.   The triviality of end-group effects in high polymers follows from the common definition: the term high polymer is reserved for long chain molecules containing at least 100 repeat units, equivalent to a relative molecular mass of about 10 000. Virtually all commercially important polymers are in this class, though properties typical of high polymers are exhibited by samples having considerably lower relative molecular masses.

As mentioned earlier a single average relative molecular mass may be inadequate to describe a polymer for a given purpose, and even two distinct averages may be a poor substitute for a complete distribution of relative molecular masses.   To determine an average or complete distribution it is usually necessary to begin by dissolving a known weight of the polymer to make a definite volume of solution. This is because the experimental techniques are based on the thermodynamics of dilute solutions, where changes in some measurable property of a solution may be related to the relative molecular mass of the solute causing the change.   Colligative properties have been studied for about 100 years, and the basic assumption is that every individual solute species, whatever its mass, has precisely the same effect on the properties of the system.   That is why a cryoscopic constant, for example, can be evaluated for a solute of known relative molecular mass and then used to determine an unknown relative molecular mass.   Even for non-polymeric systems there are familiar experimental difficulties; if the change in the measured property is intrinsically small the solution must not be too dilute or else the change cannot be measured with adequate precision.   On the other hand the simplest thermodynamic relationships become increasingly unsatisfactory with increasing concentration.   Simple theory ignores interactions between the solute particles and modification of the solvent by the solute.   When polymer molecules are involved the problem is doubly difficult.   A relatively small number of molecules, providing only a small change in colligative properties, will introduce a considerable weight of solute into the system thereby threatening its ideality as chain entanglement occurs.   In general, experiments with polymers involve the precise measurement of rather small changes, coupled with some modification of the basic thermodynamics to allow for non-ideal behaviour.   The constraints are often of a peculiarly frustrating kind: even the best solvent for a given polymer may require high temperature working where polymer degradation is possible.   The consequences of these constraints will be illustrated in the experimental sections following this general discussion of relative molecular mass measurement.

Apart from colligative properties, polymer solutions are often studied to obtain data on the size of the individual units present. The light scattering characteristics, for example, lead to the weight average relative molecular mass. Although the number and weight average relative molecular masses have a special status, they are not the only averages in common use. Viscometry is in many ways a specially suitable technique for the study of polymer solutions but the result is the viscosity average which lies between the averages mentioned above, though it is closest to the weight average.

The labour involved in obtaining a complete distribution of relative molecular masses is, even at best, quite considerable, and it is usual to make do whenever possible with one or more average values determined directly. If it should be necessary to obtain a complete distribution for a particular polymer sample it is a trivial matter to calculate any desired average from the data.

When considering experimental techniques for the measurement of relative molecular masses it is important to realize that they are not in a fixed order with respect to convenience or reliability. Improvements in the design of instruments can reshuffle the relative attractiveness of different techniques.

Experimental methods

Complete distribution of relative molecular masses. The oldest method brings out very vividly what is implied by a distribution and is a good starting point. It involves fractionation of the original polymer into a series of smaller samples or fractions whose spread of relative molecular masses is much less than that of the whole polymer. Indeed with 10 or 12 fractions it may be assumed that there is *no* spread within a given fraction; this fiction permits the use of any of the normal techniques for relative molecular mass measurement since $\bar{M}_n$ and $\bar{M}_w$ necessarily coincide for a monodisperse specimen. There is no advantage in trying to ensure that the fractions are of equal weight, rather the reverse. Relatively large fractions are acceptable in parts of the distribution where the relative molecular mass is not changing rapidly but smaller fractions are advisable towards the edges of a distribution. Any method of fractionation rests upon some property which is sufficiently sensitive to relative molecular mass to permit a physical separation in an acceptably short time. Fractional precipitation from solution is one especially well studied area. The principle is that a non-solvent is added to the polymer solution in small aliquots, and after each addition the precipitated polymer is removed. Material of low relative molecular mass is recovered last. Analysis of the separate fractions then completes the procedure which is essentially discontinuous. An obvious refinement would be to arrange for analysis to follow complete separation on a continuous basis; gel permeation chromatography (gpc) offers just this facility. A column is packed with a suitable solid support and soaked with an appropriate solvent or mixture of solvents. A dilute solution of the polymer is added to the top of the column and followed up by a steady flow of more solvent. As the solution passes down the column molecular separation occurs to an extent determined by the operating conditions, over which a considerable degree of control is possible. Polymer of low relative molecular mass is the last to appear in the emerging solution which is analysed continuously by means of extremely precise differential refractometry. Although the technique is invaluable for routine use on a large number of similar samples it is less convenient for the determination of a single distribution if this necessitates setting up a special column and other preliminaries. Calibration is an essential part of the procedure and getting the conditions right for a new polymer system may be very time consuming. The high cost of the equipment seems likely to restrict its use to laboratories where there is a steady demand for relative molecular mass distributions on a routine basis.

Older methods of continuous separation and analysis are based on the ultracentrifuge — a device which produces high gravitational fields in a sample cell embedded in a spinning rotor. When a dilute solution of a polymer is placed in this cell the molecules of different masses sediment at measurably different velocities. An optical system permits continuous observation of the sedimentation process; the actual data taking the form of profiles of refractive index against position in the cell for appropriate times. These profiles lead eventually to complete relative molecular mass distributions but the procedure is extremely laborious unless the data processing is largely automated. The ultracentrifuge may also be used to create equilibrium conditions in the cell by maintaining a relatively low gravitational field over a period of some days. The sedimentation equilibrium results from a balance between the downward thrust of the centrifugal force and the tendency towards upward diffusion. For polymer molecules of a given mass this balance will occur at a definite position in the cell associated with a particular value of the gravitational force. Each molecular species will be distributed about its own characteristic equilibrium position, and detailed analysis is possible.

Methods for $\bar{M}_n$. Raoult's law of vapour pressure lowering is the thermodynamic basis for all experimental procedures related to colligative properties of solutions. It asserts that the vapour pressure of a pure solvent is lowered by the presence of an involatile solute to an extent which depends on the number of solute particles per unit volume of solvent, but not on their nature. For a known mass of solute the number of solute particles (molecules) is inversely proportional to relative molecular mass.

Equilibrium vapour–liquid studies are notoriously difficult, even for pure solvents, and most experimental techniques for relative molecular mass determination are not concerned with the direct measurement of changes in vapour pressure. Cryoscopy and ebulliometry are familiar techniques for non-polymeric substances which may be applied to polymers whose relative molecular masses do not exceed 5000. The normal difficulty of precise temperature measurement is naturally much increased.

The measurement of osmotic pressure offers a more attractive proposition because the measured property changes very substantially. The osmotic pressure of an ideal solution may be described by an equation like that appropriate to a perfect gas:

$$\pi V = nRT \qquad\qquad 12$$

Here π is the osmotic pressure exerted by a solution of volume V at temperature T containing n moles of solute. If the mass of solute per unit volume is C and the relative molecular mass is M, then

$$\frac{n}{V} = \frac{C}{M} \qquad\qquad 13$$

Combination of equations 12 and 13 gives

$$M = \frac{RT}{\pi/C} \qquad\qquad 14$$

For a polymer it is $\bar{M}_n$ which emerges from this analysis, and the non-ideality of all but the most dilute solutions makes it necessary to plot π/C against C, and use the intercept for infinitely dilute solution as the basis for calculation of the result. The most direct manifestation of osmotic pressure which can be measured is its ability to set up a pressure differential between a solution and a solvent separated by a membrane permeable to solvent alone. The attainment of equilibrium in static osmometers of this kind tends to be both slow and difficult and the emphasis is currently on dynamic measurement of osmotic pressure. One high speed osmometer consists of a closed solvent compartment separated from the solution by a flexible diaphragm, rigidly held round its edges. As solvent diffuses through the diaphragm into the solution a strain gauge registers the deflection of the diaphragm and this may be related to the osmotic pressure. Reliable measurements are possible for relative molecular masses ranging from 5000 to 1 000 000, and the range from 20 000 to 500 000 is especially satisfactory for this technique. For relative molecular masses lower than 10 000 or so

3

the selectivity of membranes becomes suspect, while at very high relative molecular masses the osmotic pressure exerted by a solution of tolerable concentration is very small.

A most ingenious device has been developed for what is essentially vapour phase osmometry. It is suitable for relative molecular masses up to about 20 000, precisely the region in which other methods are least effective. A droplet of polymer solution is placed on a thermistor and the small temperature change caused by pure solvent vapour condensing into the droplet is measured. Calibration with appropriate standard substances provides a relationship between temperature rise and relative molecular mass.

It is sometimes possible to obtain number average relative molecular masses by end group analysis if all the groups concerned are identical and react both specifically and completely when tested with a chemical or physical probe. For example, a molecule with terminal COOH groups might be characterized by titration with alkali, but the method tends to become unsatisfactory for relative molecular masses above about 15 000 because the reactive groups become too heavily diluted with repeat units. Precisely because this method of analysis concentrates on chain ends it can yield information about chain branching when combined with a method which depends solely on molecular size.

The very largest molecules of synthetic polymers are directly visible in the electron microscope under suitable conditions, since a coiled up molecule of relative molecular mass 10^6 is likely to be of the order of 10 nm in diameter. There are some indications that relative molecular mass methods could be based on scrutiny of individual molecules by this means.

Methods for $\bar{M}_w$. The light scattering power of polymer molecules in dilute solution provides a means of obtaining much useful information about relative molecular masses and molecular configurations. Assuming that there is a refractive index difference between the particles (individual polymer molecules) and the suspension medium, light will be scattered in all directions and the scattered light will be polarized. In practice it is the light scattered at 90° to the incident beam which is used to calculate $\bar{M}_w$, but the scattering power is low and photo multipliers are necessary even when the conditions are most favourable. For relative molecular masses less than 5000 or so the light scattering is simply too feeble to be detected with instruments currently available, while at the other end of the range dust particles interfere intolerably with relative molecular masses of the order of 10^6. Very fine dust particles are difficult to remove and are indistinguishable from large molecules of polymer with respect to scattering power.

It is worth emphasizing the importance of light scattering as a method for the study of the shape of polymer molecules in solution. Rather obviously a tightly coiled molecule will behave differently from a similar molecule in a more extended conformation. It is necessary to measure the angular distribution of the scattered light intensity and the degree of polarization, together with the variation of these quantities with the wave length of the incident light.

Methods for $\bar{M}_v$. A detailed discussion of the statistical significance of this particular relative molecular mass average would be inappropriate here, and it is sufficient to note that the viscosity average became important precisely because it is easy to measure and can be related to other properties of a polymer. A dilute solution of a polymer in a solvent is more viscous than the pure solvent at the same temperature and the theory of viscometry is concerned with the relationship between this viscosity difference and the relative molecular mass of the polymer. The actual measurements involve only a viscometer, a timing device, and a well controlled thermostat. The viscometer is commonly of glass, its main feature is a length of capillary through which the passage of a known volume of solvent or solution can be timed. Individual viscometers vary a great deal in design, especially with respect to the liquid reservoirs and the arrangements for ensuring constant hydrodynamic conditions.

The following basic definitions and equations are involved in all viscometric studies.

$$\eta_r = \frac{\eta}{\eta_0} \quad \text{and} \quad \eta_{\text{sp}} = \eta_r - 1$$

where η_r, the viscosity ratio, is the ratio of η. the viscosity of a solution of concentration c, to η_0, the viscosity of the pure solvent at the same temperature and under the same hydrodynamic conditions. The symbol η_{sp} represents the specific viscosity. Finally

$$[\eta] = \underset{c \to 0}{\text{Lt}} \frac{\eta_{\text{sp}}}{c}$$

where $[\eta]$, the limiting viscosity number, is the quantity of particular interest. It measures the intrinsic capacity of the polymer molecules to increase the viscosity of the solvent and is meaningful because of its dependence on relative molecular mass. For sufficiently dilute solutions a plot of η_{sp}/c against c is a straight line, from which $[\eta]$ may be determined as the intercept at zero concentration. The relationship between $[\eta]$ and $\bar{M}_v$ is embodied in the Mark–Houwink equation, based on an earlier suggestion by Staudinger.

$$[\eta] = K\bar{M}_v^{\alpha} \qquad \qquad 15$$

In equation 15 K and α are constants for a particular polymer, solvent and temperature, and they must be evaluated by calibration before the equation can be used. The principle of calibration is to use a series of fractions whose $\bar{M}_w$ values are known from light scattering measurements. For each fraction $\bar{M}_w$ and $\bar{M}_v$ are effectively the same, and a plot of ln $[\eta]$ against ln $\bar{M}_w$ will give ln K as intercept and α as slope. Having obtained K and α for the system under investigation it is then possible to determine $\bar{M}_v$ values for unfractionated specimens of the same polymer by measuring $[\eta]$ for each such specimen. At first sight it may appear that the relative molecular mass average which results is $\bar{M}_w$ rather than $\bar{M}_v$. It must be remembered, however, that although $\bar{M}_w$ and $\bar{M}_v$ are effectively coincident for fractions they are different for whole polymer, and the point of equation 15 lies in its claim to relate a measurable viscometric property to the relative molecular mass average appropriate to that property.

There has always been much interest in polymer solution viscometry and in the relevant theory, not least because it offers a reasonably convenient means of monitoring the product from a continuously operating polymerization reactor. Experiments with polymer melts rather than solutions have produced some striking empirical relationships of the Mark–Houwink type, but much remains to be done in this difficult area.

Chain conformations

The mass of a high polymer molecule is typically several orders of magnitude larger than that of the most elaborate non-polymeric organic molecules. Nevertheless, even a polymer molecule is still a very small entity compared with the smallest sample on which one could hope to perform physical or chemical tests. A polyethene molecule with a relative molecular mass of 180 000 is only 10^{-3} mm long, a pin head could contain something of the order of 10^{15} such molecules. This is a very important fact, providing both the justification for the use of relative molecular mass averages discussed earlier and the foundation for the statistical arguments which throw so much light on the behaviour of aggregates of polymer molecules.

Within limits a polymer chain is flexible and this immediately raises questions concerning the most favoured conformations. (Rotation of a molecule about a single bond permits one *conformation* to pass into another but the *configuration* of a molecule is a fixed property. This distinction will be acceptable to most chemists but it is not recognized by all polymer scientists.) Scientific intuition, which associates high probability with disorder rather than order, tells us that a straight line or a circle are altogether less probable representations of a polymer chain than a more tangled or coiled up conformation. It has been shown that the analysis of possible conformations is an extraordinarily fruitful way of gaining insight into the behaviour of polymer molecules, both individually and in bulk. The analysis begins with the classical one dimensional random walk problem. This concerns a person taking a stated number of steps of equal length in a straight line, subject to the condition that forward and backward steps are equally probable. The starting point is the most probable finishing point, with a diminishing probability of ending up farther away in either the forward or backward direction. Suitably generalized to three dimensions this result has something to say about polymer chains and an outline of the procedure will now be given.

It can be shown that the probability, $p_x \, dx$, of ending a one dimensional random walk of N steps, each of length l, at a distance between x and $x + dx$ from the starting point, is given by the Gaussian distribution

$$p_x \, dx = \frac{\alpha}{\pi^{\frac{1}{2}}} \exp(-\alpha^2 x^2) \, dx \qquad 16$$

where

$$\alpha = \frac{1}{l(2N)^{\frac{1}{2}}}$$

By inspection of equation 16, p_x has its maximum value when $x=0$, and it goes to zero when $x = \pm \infty$. Strictly it should go to zero for $x > Nl$ or $x < Nl$ but the analysis glosses over this requirement. Equation 16 is therefore inaccurate for displacements approaching $\pm Nl$ and meaningless for larger displacements. For shorter and more probable displacements the result is satisfactory.

To extend the analysis into three dimensions it is necessary to consider the antics of an eccentric fly which proceeds in straight lines of length l but changes direction randomly at the end of each straight section. The probability of a displacement lying between x and $x + dx$ in the x direction is now given by the distribution

$$p_x \, dx = \frac{\beta}{\pi^{\frac{1}{2}}} \exp(-\beta^2 x^2) \, dx$$

where

$$\beta = \frac{1}{l}\left(\frac{3}{2N}\right)^{\frac{1}{2}}$$

Suppose that the random flights begin from the centre of coordinates. The probability that a flight ends within an element $dx \, dy \, dz$ of coordinates (x, y, z) is given by

$$p_{x,y,z} \, dx \, dy \, dz = \frac{\beta^3}{\pi^{3/2}} \exp[-\beta^2(x^2 + y^2 + z^2)] \, dx \, dy \, dz$$

It only remains to remove the restriction that the displacement shall be in a particular direction.   The result will be the probability that a random flight, consisting of a fixed number of straight line elements of constant length, shall result in a particular displacement from the starting point, irrespective of direction.   All points on a sphere, centre the origin, of radius equal to the displacement considered, will be acceptable finishing points.   Let r be the radius of such a sphere, and note that $r = (x^2 + y^2 + z^2)^{\frac{1}{2}}$.

$$p_r \, \mathrm{d}r = 4\pi r^2 \, \frac{\beta^3}{\pi^{3/2}} \exp(-\beta^2 r^2) \, \mathrm{d}r$$

$$= \frac{4\beta^3 r^2}{\pi^{\frac{1}{2}}} \exp(-\beta^2 r^2) \, \mathrm{d}r$$

The quantities p_r and r are positive, and *Fig. 6* shows that a plot of one against the other has a maximum.   (The treatment now becomes

FIG. 6.   Analysis of a random flight or freely-jointed chain consisting of 100 links each of unit length.   The probability of an end-to-end distance r is given by p_r.

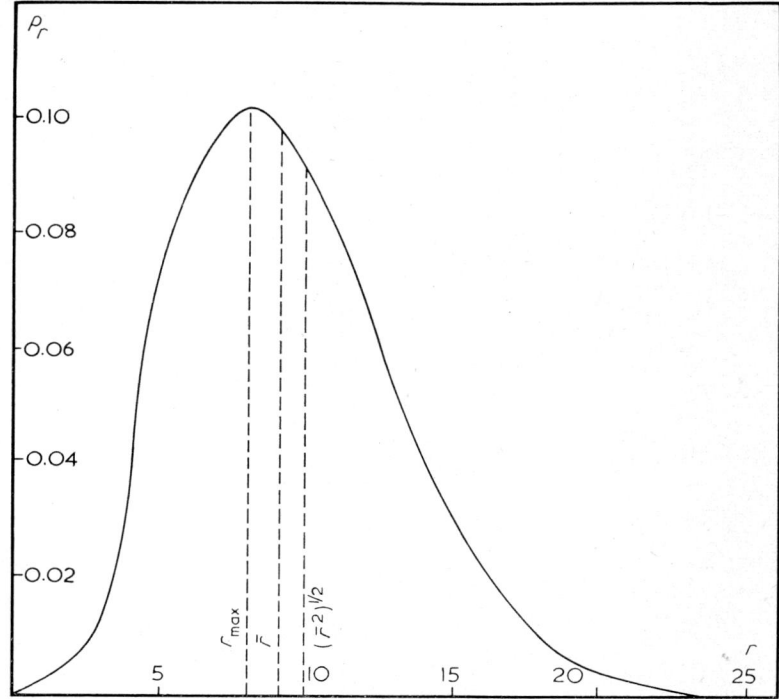

reminiscent of the $1s$ orbital problem in quantum mechanics.) Differentiation shows that $r_{max} = 1/\beta = l(2N/3)^{\frac{1}{2}}$.

By itself this result does not describe the random flight adequately. The average value of r, for example, is not quite the same as r_{max} because the curve in *Fig. 6* is not symmetrical about r_{max}. The root mean square value is also important.

$$\text{average value} = \bar{r} = l \left(\frac{8N}{3\pi} \right)^{\frac{1}{2}}$$

$$\text{rms value} = (\bar{r}^2)^{\frac{1}{2}} = lN^{\frac{1}{2}} \qquad\qquad 17$$

All of the above applies equally well to a hypothetical freely jointed chain. This useful fictional entity consists of rods of zero volume connected by joints which will accommodate all changes of direction including complete reversal. The results for r_{max}, $\bar{r}$ and $(\bar{r}^2)^{\frac{1}{2}}$ jointly characterize the innumerable conformations which such a chain could adopt. However, the freely jointed chain would appear to be somewhat inappropriate as a model for the behaviour of a real polymer chain. In the first place bond angles in a real chain are not infinitely flexible. Even for the sterically unrestricted polyethene chain the bond angles are constrained to lie within definite limits by the valence geometry of the carbon atoms. Secondly, the freely jointed chain model is based on rods of zero volume; it is quite possible for two or more links to superpose completely or to cross over one another. This matter of physical overlap is not really as serious a problem in real chains as one might think, except perhaps for very tightly coiled conformations. It is much more important to take some account of restricted bond angles and an ingenious procedure has been developed to do this.

The aim is to identify a freely jointed model chain which will reproduce certain conformational characteristics of a given real chain. An appropriate number of repeat units in the real chain are represented by a single link in a freely jointed chain, the number being large enough to ensure that consecutive links in the model chain are mutually independent (unrestricted) with respect to direction. In *Fig. 7* every sequence of five repeat units in a two-dimensional representation of a real chain is modelled by one link in a hypothetical freely jointed chain represented by the dotted lines. For the purposes of *Fig. 7* the 'real' chain is endowed arbitrarily with a constant bond angle of 120°. The links in the freely jointed model chain are not of constant length, but a root mean square value may be used in a final version of the model. In general the conformational analysis leading to r_{max}, $\bar{r}$ and $(\bar{r}^2)^{\frac{1}{2}}$ will apply to model chains with at least 10 links.

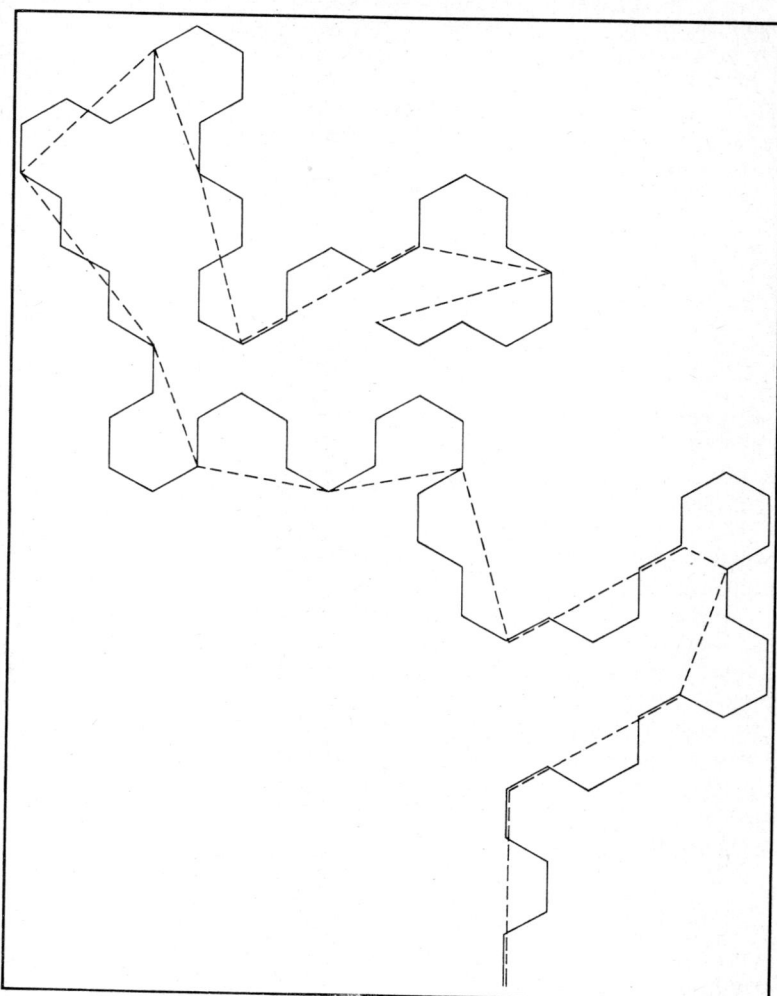

Fɪɢ. 7. Real molecular chain and equivalent model chain. The 'real' chain is shown here as a two-dimensional entity with an arbitrary angle of 120° between consecutive bonds. Crossing and overlapping are forbidden. The model chain (dotted) has one link for every five in the real chain.

If, for a given real chain containing N links of length l, the equivalent freely jointed chain has N' links of length l', then

$$(\bar{r}^2)^{\frac{1}{2}} = l'(N')^{\frac{1}{2}} \qquad\qquad 18$$

where $(\bar{r}^2)^{\frac{1}{2}}$ is the root mean square end-to-end distance for the real

molecular chain.   Equation 18 asserts the equivalence of the real
and model chains with respect to the statistical property $(\bar{r}^2)^{\frac{1}{2}}$.
It is possible to eliminate N' and l' in favour of N and l if a numerical
relationship can be established by experiment.   For polyethene

$$(\bar{r}^2)^{\frac{1}{2}} = l(2N)^{\frac{1}{2}}$$

a result which differs from that for the freely jointed chain (equation
17) by the factor $2^{\frac{1}{2}}$.   Evidently the requirement that consecutive
carbon–carbon bonds in the chain shall be at an angle of 109 ° to one
another leads to an array of possible conformations which are
somewhat less compact, and the root mean square end-to-end
distance is thereby increased.

Far from being of purely academic interest, the results of these
analyses provide an immediate glimpse of the underlying mechanism
of one of the most striking properties of certain polymers.   The
elasticity of a rubber might be expected to depend on the difference
between the length of a fully extended polymer molecule and the
dimensions of its more compact, more probable conformations.   If
a freely jointed chain containing N links of length l were fully
extended its length would be Nl.   Hence

$$\frac{\text{fully extended length}}{\text{root mean square end-to-end length}} = \frac{Nl}{lN^{\frac{1}{2}}} = N^{\frac{1}{2}}. \qquad 19$$

For 100 links, equation 19 predicts an 'extension factor' of 10, this
is certainly of the right order of magnitude for certain rubbers.   The
result would require modification for real molecules, especially as the
extended chain length will be less than the contour length because of
bond angle restrictions.   Nevertheless, the general implications of
the above ratio are correct: a piece of rubber stretches under an
applied force because the molecules abandon their original relatively
compact conformations in favour of more extended conformations.
The entropy is thereby reduced and when the rubber is allowed to
retract the molecules rapidly explore the possibilities afforded by the
new situation and re-adopt more probable conformations like those
from which they were displaced by extension.   This exploration
occurs more readily at higher temperatures which means that the
force required to produce a given extension is greater at higher
temperatures.   If unvulcanised rubber is held under stress for a
sufficiently long time the molecules gradually disentangle themselves
from one another in the extended conformations and thus gain
access to the preferred conformations.   The stress required to
maintain the extension decays to zero while this process is going on.
In practice rubbers retain their elasticity because they are lightly
vulcanised.   The cross links between chains inserted by this process

are sufficient to prevent disentanglement under stress but do not seriously reduce the overall elasticity.

The stretching of a rubbery polymer is evidently quite unlike the stretching of a metal spring. Indeed the idealized thermodynamics of the former are analogous to the behaviour of a perfect gas and it is perfectly correct to regard a piece of elastic as an 'entropy spring' since the retractive force is entropic in origin.

For a perfect gas:

$$dU = -p\,dV + T\,dS$$
$$dA = -p\,dV - S\,dT.$$

For a rubber, by analogy:

$$dU = f\,dl + T\,dS$$
$$dA = f\,dl - S\,dT. \qquad\qquad 20$$

Here l_0 is the unstretched length of a specimen, and f is the applied force corresponding to an extended length l. The terms $-p\,dV$ and $+f\,dl$ are equivalent and describe work done on the system. The signs are reversed because a perfect gas does work when it expands whereas a rubber does work when it retracts against an applied force. If the rubber is stretched at constant T and V (equivalent to assuming that Poisson's ratio is 0.5) then

$$A = U - TS$$

and

$$(\partial A/\partial l)_T = (\partial U/\partial l)_T - T(\partial S/\partial l)_T. \qquad\qquad 21$$

For a perfect gas $(\partial U/\partial V)_T = 0$, and by analogy $(\partial U/\partial l)_T = 0$.

Also $f = (\partial A/\partial l)_T$ from equation 20, hence equation 21 becomes

$$f = -T(\partial S/\partial l)_T \qquad\qquad 22$$

This remarkably simple result is the justification for bringing thermodynamics into a section concerned with the conformations of polymer molecules. It expresses quantitatively the conclusion reached qualitatively in the earlier part of the discussion by a wholly different kind of argument. Equation 22 shows that the force required to produce a given extension in a rubbery specimen depends on the extent to which the entropy changes with extension. Rather obviously $(\partial S/\partial l)_T$ is intrinsically negative, and it may be shown that it is actually independent of T. The expression $f = kT$ is therefore an acceptable description of the entropy spring, just as $p = (R/V)T$ is appropriate for a perfect gas. The precise identity of k is easily revealed as the product of two familiar physical quantities which can be measured directly.

$$k = -(\partial S/\partial l)_T = (\partial f/\partial T)_l = -(\partial l/\partial T)_f(\partial f/\partial l)_T = \left\{-\frac{1}{l}(\partial l/\partial T)_f\right\}\{l(\partial f/\partial l)_T\}$$

= coefficient of linear expansion × elastic modulus.

In summary it may be said that the freely jointed chain model provides a vivid insight into the manner in which the behaviour of polymer molecules is controlled by the availability of alternative conformations of varying probability. Unfortunately the model is of no value whatever when the properties of individual segments are considered in a realistic light. The model does no more than match certain specified statistical properties of the real chain. Several distinguished workers are currently developing better descriptions of real chains, taking proper account of short range forces and other restrictions. An account of their methods cannot be given here, where the freely jointed chain model must stand alone as a crude example of statistical analysis applied to long chains.

4. Aggregates of Polymer Molecules

General features

The description of polymeric solids, liquids and solutions, and of transitions between them, involves a set of contrasting ideas concerned with motion and immobility, disorder and order, entanglement and segregation. In ordinary molecular, atomic or ionic materials of a non-polymeric kind, interest centres on the interactions between the discrete units and how they position themselves relative to one another. The theory of liquids involves nearest neighbour patterns, while the solid state, if crystalline, demands a very precise organization of the individual units. In a polymer much of this remains true, the repeat units may be identified with the structural units mentioned above. There is, however, an important new feature. Among the nearest neighbours of a given repeat unit there will be other units, commonly two, which are joined to it by covalent bonds. The force between units so connected is therefore much greater than that between adjacent but unbonded units. The result is that a quite novel kind of constraint operates in polymer systems; a given repeat unit may not move from its initial position without bond rupture unless its neighbours in the chain also move to new positions which are compatible with the length and direction of the valence bonds. These neighbours will have other neighbours and so on down the chain. Considerations of chain continuity will also influence the packing of polymer molecules in the solid.

The main consequence of this structural peculiarity is that polymers exhibit a wide range of properties depending on the temperature which controls the mobility of the chains. The transition from one distinct state of aggregation to another is commonly more complicated than a phase change in a non-polymeric material. An understanding of chain mobility and chain conformations provides a basis for the description of transitions.

It is not easy to decide upon the best order in which to treat glasses, rubbers and semi-crystalline polymers. The approach adopted here is designed to emphasize the intermediate position of semi-crystalline polymers and to highlight the complementary properties of the glassy and rubbery states.

Semi-crystalline polymers

In the late 1920s electron diffraction techniques were applied to a number of natural polymers, notably cellulose. It was shown that

FIG. 8. Two-dimensional representation of the Fringed Micelle Model for the structure of a semi-crystalline polymer. The crystalline regions or crystallites are embedded in, and connected to, the non-crystalline matrix.

there were definite regions or crystallites where the polymer repeat units were spatially organized in a regular manner corresponding to a unit cell. Not unreasonably it was assumed that each crystallite consisted of bundles of molecules arranged parallel to one another. The length of a crystallite, commonly ranging from a few tens to a few hundred nanometres, therefore defined an upper limit for the length of the molecules concerned. Largely as a reaction to this view, another model was developed in which a given polymer chain was assumed to exist partly in one or several crystallites and partly in the intercrystalline material, its length now being unrelated to the dimensions of the crystallites. This was the Fringed Micelle Model which held sway for the best part of 30 years. A planar representation of the model appears in *Fig. 8* but it must be remembered that the structure is really three-dimensional. In its turn this model has been largely supplanted but it will serve for the purposes of this

section.   Newer views will be mentioned later in the light of develop-
ments in another area of polymer science.

The presence of crystallinity in polymers remains somewhat
surprising.   Wholly crystalline specimens are inconceivable because
even the purest polymers are heterogeneous to some degree.   The
heterogeneity associated with chain ends in particular is a guarantee
that the chains will not secure complete integration into a structure
based exclusively on repetition of the unit cell.   The kind of crystals
formed by sugar and salt have no structural counterparts in the world
of polymers, and polycrystalline structures, associated with metals,
are also absent.   Instead, a polymer which is not wholly amorphous
will have some intermediate state of order.   This intermediate state
of semi-crystallinity is the most important single feature of many
of the most important synthetic polymers such as polyethene,
polypropene and nylon.   An enormous amount of effort has been
applied to the problem of understanding the structure of these and
other crystalline polymers.   (The qualifying prefix semi- is very
often omitted.)   The proportion of a sample in the form of crystallites, commonly on a weight basis, is known as the degree of crystallinity, or simply the crystallinity.   This definition is not as precise
as it might appear but crystallinities in excess of 85 per cent are rare
on any basis.   The fringed micelle model and the concept of crystallinity lead jointly to the view that a semi-crystalline polymer is a two
phase system.   For many purposes this serves very well but the term
phase is not being used in the strict sense of the Phase Rule.   In
particular the physical properties of the non-crystalline 'phase' will
not be precisely the same at all points though overall it may respond
well to tests of its homogeneity.

To make practical use of the two phase model it is necessary to
attribute average properties to each phase.   Any overall property
may then be expressed in terms of the contributions from each phase,
weighted according to the degree of crystallinity.

$$P = w_1 P_1 + (1 - w_1) P_2 \qquad\qquad 23$$

In this equation P is some overall property per unit mass, w_1 is the
weight fraction of phase 1, the crystalline phase, and P_1, P_2 correspond to the values of the property per unit mass of pure phases 1
and 2 respectively.   This equation underlies all assessments of the
crystallinity and it is hardly surprising that different experimental
methods lead to somewhat different results for the same polymer
specimen.   One can never be quite sure if the pure phase properties,
however carefully measured, are always characteristic of the phases
concerned whatever their relative proportions.   This is equivalent
to saying that the distinction, if any, between partial specific proper-

ties (which depend on composition) and pure phase properties (as used in equation 23 above) is not clear. Among the properties commonly measured are density, specific heat, specific enthalpy, heat of fusion, ir spectrum and nmr line widths. The crystallinity based on density is especially common and convenient and will be used to illustrate the method.

The density of the specimen as a whole can be measured with high precision in a suitable density gradient column. This is a thermo-statically controlled column of liquid whose composition and density vary linearly from top to bottom. Calibration with floats of known density provides a graph of density as a function of position in the column because each float comes to rest where its density is matched by that of the ambient fluid. A small specimen of the polymer, dropped gently into the column, comes to equilibrium in a relatively short time and its position is noted. The density of a perfectly crystalline phase may be calculated from the dimensions of the unit cell as revealed by x-ray crystallography. Unfortunately, the density of the non-crystalline phase is more difficult to determine. Polyethene, for example, cannot be prepared in the solid form without a notable degree of crystallinity and it is necessary to use data for the melt in the temperature range 150 °C to 200 °C. These data are extrapolated to give the density at ambient temperatures which, in effect, refers to a supercooled melt. Subject always to the intrinsic limitations of the two phase model, the following expressions for the crystallinity apply:

$$w_1 = \frac{\rho_c(\rho - \rho_a)}{\rho(\rho_c - \rho_a)} = \frac{v_a - v}{v_a - v_c} \qquad 24$$

In these expressions, based on equation 23, ρ and v represent density and specific volume, while the subscripts c and a conventionally denote crystalline and amorphous material. The absence of a subscript indicates an overall value characteristic of the complete specimen.

Having noted the presence of a crystalline phase in some polymers it is appropriate to consider the melting of crystallites in terms of the thermodynamics of simple first order transitions. For a given pressure, the melting point of a crystalline solid is defined as that temperature at which the pure solid can coexist with the pure melt. The transition from one phase to the other occurs, in theory, at a perfectly definite temperature and there are step changes in the primary thermodynamic properties such as density, enthalpy and entropy. The Clausius–Clapeyron equation describes the melting process

$$\frac{dT}{dP} = \frac{T\Delta V}{\Delta H}. \qquad 25$$

The change of melting point, T, with pressure, P, is specifically related to the step changes in two properties, the enthalpy, H, and the volume, V.

At first sight the melting of a crystalline polymer would appear to differ qualitatively from the kind of transition described above. *Figure 9* shows a typical plot of specific volume against temperature for such a polymer; the melting is spread over $10\,°C$, even when the experiment is carried out very slowly. The final melting temperature is sharp to $0.5\,°C$ or better and most of the specimen may melt in its vicinity. Given the temperature dependence of the quantities involved, equation 25 would show that the crystallinity of the specimen falls off as the melting proceeds, with a sharp fall towards the end of the process. The problem is to explain how the melting process can be diffuse and yet remain a first order transition. The explanation lies in the fact that the surface free energy of a crystallite is by no means negligible, unlike the surface free energy of a visible crystal of a non-polymeric solid. Melting involves the elimination of this energy and the melting point is thereby lowered, especially for the smaller crystallites, below the value appropriate to a large crystal of negligible surface free energy. The intrinsic heterogeneity

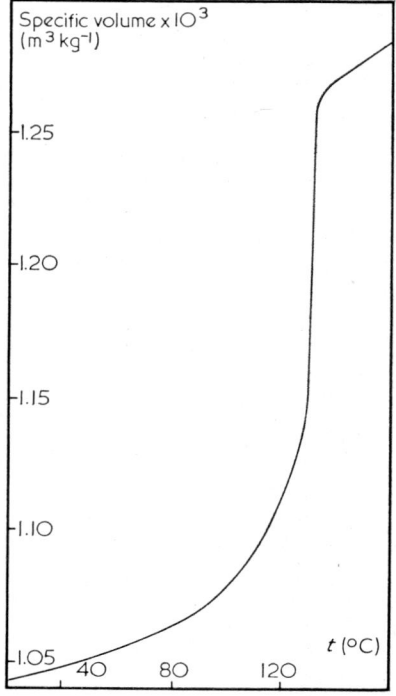

FIG. 9. Melting of a semi-crystalline polymer. Change of specific volume with temperature in a typical dilatometric experiment.

of any polymer will inevitably produce a range of crystallite sizes and hence a range of melting points, but the thermodynamic criterion for a first order transition is met by each individual crystallite. There remains the formal definition of the melting point of a crystalline polymer and it seems best to identify the highest temperature to which crystallinity can survive. This melting point will be lower than the true thermodynamic melting point which applies to chains of infinite length but a detailed analysis of this complication will not be attempted here.

The application of macroscopic thermodynamics to melting leads to some useful generalizations. At the melting point, T_m,

$$\Delta G_m = 0 = \Delta H_m - T_m \Delta S_m$$

whence $T_m = \Delta H_m / \Delta S_m$. This expression for T_m predicts that a high melting point can be the result of a high value of ΔH_m and/or a small value of ΔS_m. The former corresponds to especially firm binding of adjacent but unbonded units in the polymer lattice. Polymers with compact repeat units capable of close approach to neighbours in other chains are in this category, also polymers with some capacity for specific interactions such as hydrogen bonding. If ΔS_m is small, melting does not result in a large gain of conformational entropy and, to some degree, the structure of the solid must persist in the melt. Molecules of isotactic polypropene crystallize in the form of helices and these are thought to occur in the melt also, thus making ΔS_m small and T_m high.

Crystallization is the reverse of melting, and is a process of great practical importance. Polymer processing normally involves the melt from which the desired articles are fabricated by a suitable technique combining shaping with cooling. The rate and manner of cooling controls the texture of the product if the polymer is crystalline and this in turn has a bearing on the mechanical and other properties of the object produced. The crystallization of polymers is also a matter of considerable academic interest since it involves the ordering of long chain molecules. It has proved quite exceptionally difficult to account for this ordering in terms of molecular processes. The Avrami equation, which has provided the starting point for many studies of polymer crystallization, is itself non-molecular. It was originally proposed in the general context of phase changes, and has been adapted, perhaps too enthusiastically, to describe the crystallization of polymers. At least its shortcomings indicate where more work is required. It relates the fraction of a sample still molten, θ, to the time, t, which has elapsed since crystallization began. The temperature must be held constant.

$$\theta = \exp(-Zt^n) \qquad\qquad 26$$

4

For a given system under specified conditions Z and n are constants and, in theory, they provide information about the nature of the crystallization process. Taking logs twice in succession gives

$$\ln(-\ln\theta) = \ln Z + n \ln t.$$

A plot of $\ln(-\ln\theta)$ against $\ln t$ should be a straight line of slope n, making an intercept of $\ln Z$ with the vertical axis. In practice this method of evaluating n and Z is very prone to error and curve fitting procedures using equation 26 and the raw data are normally preferred.

Like any other crystallizing system, molten polymer will solidify when nuclei are being generated in sufficient numbers. The homogeneous production of nuclei requires a considerable degree of supercooling but adventitious nuclei become potent at modest degrees of supercooling and are normally responsible for the observed phase change. Once a nucleus has been formed it may develop in a number of different ways but spherical growth is the most common, the spherulite is therefore the most familiar gross unit of crystallinity in polymers. Detailed morphological studies have shown that it is not at all like an onion but consists of fibrils which radiate outwards from the centre and branch frequently so as to occupy almost all of the space available. It is generally thought that the gaps between the fibrils are filled with largely uncrystallized or even uncrystallizable material, though some further crystallization may occur very slowly in these regions after the primary crystallization is virtually complete. The spherulite is such a complicated structure that models for its growth tend to be oversimplified and based on a smooth sphere of definite age whose radius increases linearly with time. The exponent of the Avrami equation, n, has a theoretical value of three when crystallization takes the form of spherulitic growth of nuclei which came into being at the same instant in time. Integral values of n ranging from one to four can be attributed to other forms of nucleation and growth. A typical investigation might involve the measurement of the overall rate of solidification at a number of temperatures, backed up by microscopic observation of nucleus formation and growth. In favourable circumstances, application of the Avrami equation to the former data will lead to conclusions compatible with the direct observations. The actual rate is commonly measured by dilatometry, where the change in specific volume on solidification is followed by noting the level of the top of a column of mercury in a capillary connected to the chamber containing the polymer sample in a thermostat. The height of the column is directly proportional to specific volume hence, from equation 24

$$1 - \theta = \frac{v_0 - v_t}{v_0 - v_\infty} = \frac{h_0 - h_t}{h_0 - h_\infty} \qquad\qquad 27$$

where h is height and the subscripts indicate the time which has elapsed since the beginning of the experiment. As before v is specific volume and there is naturally some kinship between equations 24 and 25. Note, however, that w_1 was the weight fraction of material in the (perfectly) crystalline form, whereas $1 - \theta$ is the weight fraction of *solid* material.

Even when a crystallization is studied over a range of temperatures so that the temperature coefficient of Z may be evaluated, it is difficult to read very much into the results. The Avrami equation is at best a reasonable overall description of the crystallization process and the separate stages of nucleation and growth have separate activation energies which are not distinguished by the Avrami analysis. Recently, the whole framework of the classical approach to nucleation and growth has been called into question, largely on the grounds that it involves the application of broad thermodynamic principles to systems which are, initially, quite small in terms of the number of molecules present. Nucleation and growth are the results of blind chaotic molecular motions taking place under the influence of driving forces associated with super-cooling. The neat classical models are inevitably a poor representation of these processes and it is now becoming possible to simulate crystallization by computer methods and make some proper allowance for features ignored by the classical arguments. This approach seems very likely to throw some much needed new light on the whole question of phase change.

Polymer single crystals and their relationship to polymer crystallized in bulk

Although the existence of some degree of crystallinity in many bulk polymers has been recognized for almost half a century, single crystals of polymeric substances were not identified and characterized until relatively recently. Only with electron microscopes of sufficient resolving power is it possible to study the very small crystalline entities obtained from dilute solutions of polymers in suitable solvents. This branch of polymer science is commonly dated from 1957, though some important preliminary work was reported earlier. From the beginning, polyethene has figured very prominently in single crystal research, but many of the conclusions reached may be of quite general applicability.

In 1957 it was a matter for considerable surprise that polyethene could be crystallized from dilute dimethyl benzene (xylene) solution to give regular crystals with distinctive features. The range of chain

4*

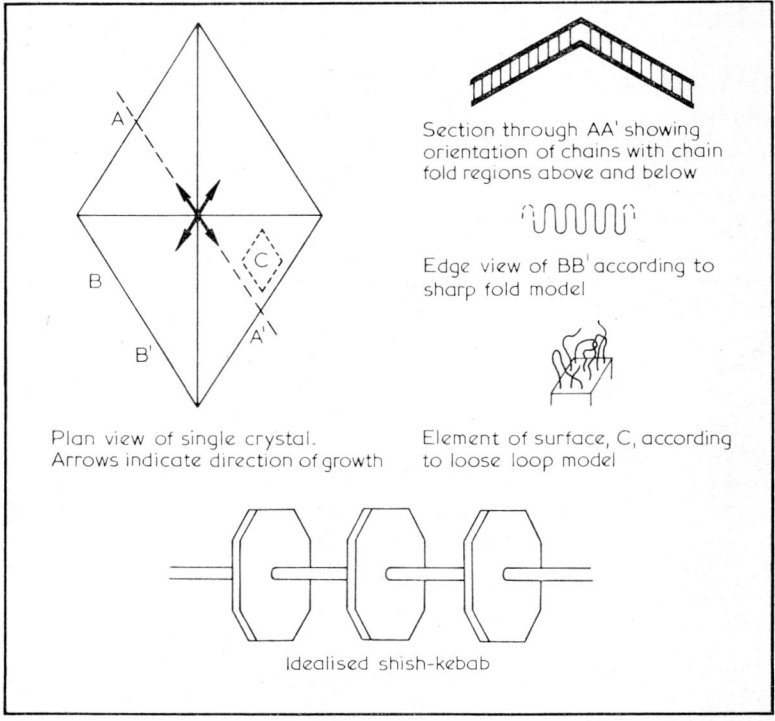

Plan view of single crystal. Arrows indicate direction of growth

Section through AA' showing orientation of chains with chain fold regions above and below

Edge view of BB' according to sharp fold model

Element of surface, C, according to loose loop model

Idealised shish-kebab

FIG. 10. Some features of polymer crystals prepared from dilute solution and theories concerning their structure.

lengths in the samples used then and since makes it hard to understand how the molecules cooperate to produce an entity with overall regularity, and not merely the localized unit cell regularity encompassed by the hitherto unchallenged Fringed Micelle Model for the texture of semi-crystalline polymer. *Figure 10* illustrates some features of polyethene single crystals. They are typically a few μm in length, but the thickness is of the order of only 10 nm. Although lozenge shaped when viewed from above, the crystals are hollow pyramids whose lower edges do not lie in one plane. Special crystallization procedures are necessary to obtain highly uniform crystals which are both large and free from overgrowths but straightforward crystallization from 1,4 dimethyl benzene at 85 °C produces representative forms if the concentration is kept below 0.01 per cent. The overall regularity of the best crystals is remarkable enough but the polymer chains are arranged in a wholly unexpected manner. The chain axes proved to be parallel to the main axis of the pyramid.

Now since the average chain length is commonly an order of magnitude greater than the crystal thickness in the chain direction, it is necessary to conclude that the molecules somehow fold back and forth upon themselves, any given molecule making a number of passes through the crystal.   Precisely how this is achieved remains uncertain, even though chain folding was proposed in 1938 for certain gutta percha films and confirmed for polyethene crystals in 1957. The fold length, or distance between folds, depends on the temperature of crystallization but is independent of relative molecular mass provided the chains are long enough to make at least one complete pass through the crystal.   This is why molecules of very different lengths can be accommodated in the same regular crystal.   At present it appears that the dependence of fold length on temperature demands an explanation in terms of the kinetics of crystal growth, though a thermodynamic theory was popular at one time.

Chain folding is now known to be closely connected with another curious property of single crystals.   The density of polyethene in this form is 3 per cent below that of a 'perfect' crystal as indicated by the dimensions of the unit cell.   There are good grounds for thinking that each crystal is actually a two-phase system in which up to 83 per cent of the material does have the unit cell density, while the remainder has a considerably lower density (1000 and 850 kg/m^3 respectively).   The idea of an apparently smooth regular crystal containing a notable amount of intrinsically non-crystalline material is startlingly unfamiliar but this material may be associated with the fold surfaces or fold planes.   These occur on the flat upper and lower surfaces of the crystals where the chains fold to effect a change of direction after each passage through the crystalline 'core'.   In addition, chain ends may be concentrated in these regions thus contributing to the lowered density.   This density has been commonly identified with that of wholly amorphous polymer, though there are certainly differences between the non-crystalline regions in bulk crystallized polymer and the fold planes in single crystals.

The Sharp Fold Model (*see Fig. 10*) accounts for the overall regularity of the crystals, including their pyramidal morphology, but it fails to explain why the surface planes are of lowered density. Tight, compact folds are no less dense than the core itself.   The so-called Switchboard Model has complementary merits and defects. Instead of chains folding neatly and re-entering the crystal at points adjacent to those from which they emerge, this model proposes a surface region constituted by a jumble of loose loops and chain ends after the appearance of a telephone switchboard.   Certainly this gives a very plausible account of the density deficiency associated with the surface regions but it does not explain the characteristic morphology.   A compromise model asserts that the pyramidal morphology follows directly from the spatial requirements of the chain folds and that the folds, though tidily organized, are just slack enough to constitute the low density surface.

Under conditions of brisk stirring, crystallization from solution gives fascinating 'Shish-Kebab' structures, in which platelets of folded chain crystals are strung along a central filament of extended chains. When first discovered this appeared to be merely an academic curiosity but it has been established that similar structures occur in polymer crystallized from the melt, especially under shear, and even in nascent polymer as formed at the surface of a solid catalyst.

The enormous amount of work on single crystals has thrown some light on the more complicated task of characterizing the formation, structure and properties of ordinary bulk crystallized polymer in adequate molecular detail. Interesting results were reported in 1963 concerning the digestion of polyethene with fuming nitric acid. Under appropriate conditions the non-crystalline regions were preferentially destroyed, and the debris proved to contain lamellar fragments very reminiscent of the crystals obtained from dilute solution. It now seems certain that chain folded entities are indeed the fundamental units of crystallinity in bulk crystallized polymer, though they tend to be somewhat disguised and incorporated in larger units. The spherulite, as mentioned earlier, consists of fibrils radiating outwards from the centre and it appears that the fibrils are constituted by packets of superposed lamellar crystals, joined together packet by packet. The polymer chain axes are normal to radii of the spherulite. If the spherulite is regarded as the most complicated crystalline entity with the single crystal as the simplest, it is natural to wonder if units of intermediate complexity can be generated under appropriate conditions. Crystallization from more concentrated solutions does indeed yield more complicated crystals and a hierarchical progression has been proposed.

The unification of existing knowledge of single crystals and bulk crystallized polymer is a matter of current interest. For example, appropriate quantities of single crystals may be agglomerated to form macroscopic specimens whose properties are measurable by normal methods. Much useful information may be gleaned from the behaviour of such material, especially by the controlled use of annealing. It is a general property of folded chains that they will refold to longer fold lengths on annealing at a temperature above the original crystallization temperature. An agglomerate of initially discrete single crystals acquires a strong interlocked structure if it is suitably annealed.

The effect of high pressure on the crystallization and annealing of polymers has been much studied in recent years. It appears that very thick crystals with few folds are produced, commonly known as extended chain crystals. Specimens of this material are very dense and brittle. Similar material can be produced by prolonged annealing under vacuum at progressively increasing temperatures and this suggests that pressure merely accelerates an intrinsic process which occurs under annealing conditions. Crystallization from the melt under pressure might well be a distinct process.

Amorphous polymers

Molten polymers

The adjective amorphous means that there is an absence of crystalline order and a structure so described is indefinite. Apart from those crystalline polymers which maintain some conformational order on melting, it is appropriate to describe molten polymers as amorphous and the model provided by a bowl of spaghetti is quite helpful though incurably inanimate. A rather special mechanism is required to explain bulk flow by a molten polymer because it is hardly credible that a whole molecule gathers itself together and jumps bodily to a new position. In a simple liquid the molecules tumble over and over one another and the observed viscosity of the system is due to the intermolecular forces which in turn largely depend on the average distance of separation. (A curious and interesting consequence of this is that viscosity is not very dependent on temperature if it is measured at constant *volume*. The familiar decrease in viscosity with increasing temperature is mainly due to the expansion which occurs in experiments at constant *pressure*.) In a polymer molecule a given segment is 'aware' of its immediate neighbours in a very intimate way, the motion appropriate to bulk flow involves some degree of mutual cooperation between segments. If rotation occurs about the marked bonds in *Fig. 11*, a number of segments are displaced collectively to the new positions indicated by the dotted lines. Runs of more than about 30 segments are rarely involved, since there are always other chains in the vicinity to contend with, all moving and jostling in a confused reaction to the applied stress causing flow. Not surprisingly a cross-linked polymer will not flow, and basically the same considerations explain why it will not crystallize either. There is not enough scope for manoeuvre by sections of appropriate length. Limited mobility, compared with the melt, is the key to understanding two other important states of aggregation shown by essentially amorphous polymers.

Rubbers and glasses

It is perhaps unusual to begin a general discussion by citing a specific and untypical material but it happens to provide a route to some

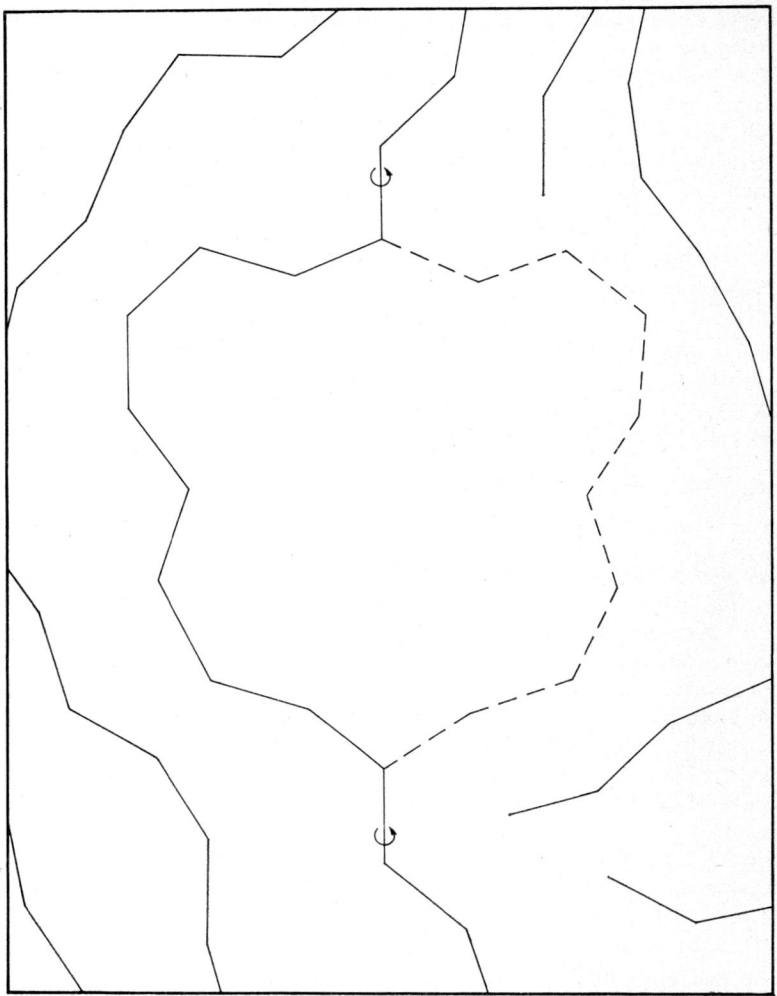

FIG. 11.   Flow in bulk polymer.   Simultaneous displacement of several repeat units by fortuitous rotation about two bonds.   Original position in solid lines, new position dotted.

general observations.   Unvulcanised natural rubber poly(*cis* methyl-buta-1,3-diene) (poly-*cis*-isoprene) is stereochemically very regular, yet it is a classical example of a substance showing very high elasticity. The capacity for crystallization, which one would expect in so regular a molecule, is made manifest by holding a sample for some time at say 0 °C, particularly in a stretched condition.   In these circum-stances it will slowly crystallize to give a stiff, hard, non-brittle material quite unlike the glassy structure normally associated with rubbers at low temperatures. The actual rate of crystallization is very slow compared with the normal semi-crystalline polymers considered earlier and this is a crucial difference. If natural rubber is cooled fairly quickly through the temperature range where crystallization *could* occur, crystallization will *not* occur. Instead, at a lower temperature, the so-called glass transition will take place to give a characteristically brittle solid. A description of this transition will be followed by further comments on glasses and rubbers.

The glass transition is not a first order transition and the primary thermodynamic quantities such as H and V do not undergo a step change. Instead, the first derivatives of these quantities with respect to temperature (the specific heat and the coefficient of cubical expansion) show the step change. Unlike these fairly sharp changes the elastic properties of the material change more gradually around the glass transition temperature, T_g, and a transition range of 50 °C is not uncommon. Rubbery properties become increasingly prominent above the T_g, while below it the material becomes increasingly glassy. Unfortunately the glass transition and its temperature are not without ambiguity, both in theory and in practice. The theory will not be discussed here but the value of T_g depends on the rate at which the experimental determination is carried out. Striking a polymer specimen with a hammer and extending it slowly would not necessarily lead to the same conclusion about its physical state.

In practice the T_g of a polymer effectively defines a temperature above or below which it may not be used. Design consideration will specify a rubber or a glassy polymer for a particular application and not a material liable to transform from one to the other under normal working conditions. For most purposes it is the position of T_g with respect to ambient temperature which matters. A rubber is therefore a substance with a relatively low T_g while a glassy polymer has a high T_g. Polybutadiene, styrene–butadiene copolymer and butyl rubber are 'rubbers' in this sense, while polystyrene, pvc and polymethylmethacrylate are 'glassy'. Even for glassy polymers behaviour above the T_g is of great practical importance because processing is normally carried out within 100 °C of the T_g and it is necessary to have some understanding of the viscosity and other relaxational properties of the material. An empirical correlation of properties with temperature is embodied in the WLF equation, named after its inventors, Williams, Landel and Ferry.

$$\ln \frac{\eta_T}{\eta_{Tg}} = \frac{-a(T - T_g)}{b + (T - T_g)}$$

Here T is a temperature above T_g, while a and b are empirical and largely universal constants. The ratio of the viscosity at T to the viscosity at T_g is given by η_T/η_{Tg}. This simple equation can give a sensible account of the behaviour of a polymer over the range from T_g to $T_g + 100\,°C$, the same numerical values of the constants applying to many polymers and to some non-polymeric materials as well. To establish the physical significance of the constants a and b it is necessary to devise a formal 'derivation' of what was and remains an empirical equation. It seems that the free volume and the coefficient of cubical expansion are somehow involved but the last word has not been said. Empirical equations of this kind are a standing challenge to the theoretician but their predictive power is valuable to the plastics engineer meantime.

Clearly it is the degree of mobility in a given polymer specimen which determines its overall properties at a stipulated temperature and whether it should be described as rubbery or glassy. A glass is a material which has lost most of its mobility but which has not otherwise changed on cooling. On the basis of x-ray diffraction, for example, a glass is indistinguishable from the corresponding rubber on the other side of the glass transition. Broad haloes characteristic of amorphous material are observed at all temperatures. The lack of mobility in a glass is naturally connected with the appearance of forces opposing mobility. At very low strains, certainly less than 1 per cent and perhaps much lower, glasses are perfectly elastic. A displacement will set up restoring forces which, on removal of the stress, bring virtually every atom back to its original position. The catastrophic fracture of glasses subject to somewhat larger strains is due to the fact that the restoring mechanism soon fails and complete breakdown of the structure follows. The glassy state may be regarded as the alternative to the crystalline state. Natural rubber, as mentioned above, can gain access to either state depending on the conditions but this is unusual. Most polymers which completely fail to crystallize in bulk are intrinsically incapable of so doing, however favourable the conditions of cooling. This is due to their innate molecular dissymmetry or segmental bulk, which precludes the close regular packing appropriate to a crystal.

When a glassy polymer is heated, the gradual appearance of elasticity to a marked degree is evidence that the molecules are becoming increasingly capable of undertaking the rapid conformational explorations on which this kind of elasticity depends. The discussion of high elasticity in the section concerned with chain conformations should now be recalled (p 37). The forces between non-bonded repeat units must be as weak as possible to facilitate the chain slippage necessary for substantial extension of the specimen. In the absence of cross links inserted by vulcanisation, molecules extended under stress will gradually disentangle themselves to relieve the stress, the deformation thereby becoming irreversible. This capability is offset by cross links, though at the price of limiting the elasticity to some degree. It is precisely because of variations in the amount of cross-linking that motor tyres and elastic bands can be so different and yet belong to the general class of materials known as rubbers. It is important not to be confused about the physical meaning of elasticity. Because of the ubiquitous elastic band, it is easy to associate this property of extensibility only with a very limited class of rubbery materials. However, all materials, even glasses, are elastic up to a point. What distinguishes a highly elastic material like an elastic band is its capacity for very long range elasticity which depends on the molecular mechanism peculiar to long chains discussed earlier. Even in these remarkable materials only a very little of the total possible extension is perfectly Hookean, with the strain proportional to the stress.

While it is true to say that semi-crystalline polymers are character- ized by a melting point and amorphous polymers display a glass transition, there is actually some duality in materials of the former type. Semi-crystalline polymers contain some non-crystalline material and in consequence undergo a glass transition, the feeble- ness of which is naturally related to the overall crystallinity. For such polymers the T_g is normally well below ambient temperatures and the transition itself is of no great practical importance. What is significant is that a semi-crystalline polymer above its T_g and below its T_m has rubbery non-crystalline regions. This explains why it falls between glassy polymers and rubbers in terms of certain physical properties. Being tougher than a glass it is much less brittle, and being stiffer than a rubber it is much less easily extended. Under appropriately large stresses it will exhibit considerable plasticity, as described in the next section. The difference between elastic and plastic behaviour is profound. The former represents a firm preference for a return to something resembling the original state after deformation, while the latter manifests a willingness to adopt a sequence of irreversibly attained states in response to an external stress.

Useful properties of solid polymers

This section features a discussion of some mechanical properties since these largely determine the use of polymerics as engineering materials. A few of the many other important properties will be mentioned first and it is especially noteworthy that the cost of a polymer is a key property which must be acceptable for any given application. All other properties will be jointly determined by the chemical constitution and structure of the polymer chains, the relative molecular mass and its distribution, and the type of aggregation, together with property modifications brought about by the conditions of processing or the use of additives.

To characterize a new thermoplastic as a potentially useful material involves much more than the physicochemical measurements associated with an academic study of average chain properties. For a given application *some* properties will be unimportant but in general it will be necessary to ensure that a whole range of properties lie within acceptable limits under service conditions.

Some properties are less fundamental than others but this does not make them any less important. The density and electrical resistance of an isotropic material are definite intrinsic properties which can be expressed in an unambiguous numerical form. The result is independent of the size and shape of the test specimen and of the experimental method. This cannot be said of properties like impact strength or resistance to cracking since they require definition in terms of a specimen of fixed dimensions and a particular test routine. In general it is quite pointless to discriminate between properties on the basis of how 'fundamental' they are.

It is largely the chemical constitution of a polymer which controls its resistance to acids, alkalis, solvents, oils and greases. Even when no chemical attack occurs, intolerable physical changes may be produced. The absorption of a solvent will swell a specimen and distort its shape. General principles have been established and they are very useful because a good deal may be inferred simply from the extent to which a material absorbs water and resists oxidation by air or oxygen. Continuous exposure of a polymer-based article to all weathers demands an appropriate resistance to the formidable group of reagents provided by sunlight, air, water and pollution. In the context of weathering it often happens that the appearance of the product may become unacceptable long before any other properties have changed significantly. Polymers normally contain specially formulated and largely secret stabilizing additives which help to maintain chemical stability. In recent times there has also been some interest in the opposite problem, that of endowing a plastic with a definite life span and a built-in capacity for self-destruction on exposure to bright light.

All properties, both physical and chemical, are temperature dependent and a given polymer will be useful only within definite ranges of temperature. Obviously the melting point or softening point must not be exceeded, exposure to boiling water or a domestic iron are examples of everyday situations which must be allowed for by the designer of polymer based articles. The glass transition, if prominent, must not occur in the working range of the product, though much can be done to reduce a T_g by the use of plasticizers. The flammability of a polymer is commonly less than that of wood or cotton but requires careful testing, especially if toxic vapours or dense smoke may be produced.

The appearance and optical properties of a plastic article are very important to the consumer. The transparency and sparkle of a wrapping film, for example, virtually determines its usefulness, given acceptable mechanical properties. The hardness of common plastics is much less than that of metals or ceramics but for many purposes their resistance to chipping or indentation is adequate. With specific gravities not far from unity, polymers are attractively light and particularly suitable for space-filling applications where no great strength is required. Many polymers are excellent insulators for electrical applications.

A polymer must not only have acceptable properties in service as part of a finished product but the actual production of the article from the raw materials must be technically acceptable, essentially on a cost basis. The material must have tolerable processing characteristics at temperatures well above those which will be encountered in service. Properties such as thermal conductivity, coefficient of expansion and viscosity have a bearing on the kind of equipment which may be used for processing, the conditions of which may determine some of the necessary additives.

Mechanical properties

The mechanical properties of both crystalline and amorphous polymers are very important and involve effects directly attributable to the macromolecular constitution of polymers. Apart from the familiar elastic moduli, the toughness, creep and fatigue characteristics of polymers are measured on a routine basis. The relative molecular mass is specially important in situations where large deformations are involved, as in flow or fracture, while the molecular structure, which controls interchain bonding, may be crucial for the smaller deformations which occur during transitions and when moduli are measured at low strains.

What makes the mechanical behaviour of polymers both interesting and complicated is the fact that they are viscoelastic. If a simple viscous liquid is stirred or deformed in some way, the energy is dissipated as heat because there is no way of storing it in the system. This contrasts with the idealized behaviour of a piece of elastic where the energy required to stretch it can be stored and recovered. A polymer partakes of both of these properties to a variable extent depending on the temperature and on its own constitution. At ambient temperatures some polymers are elastomers, some are tough solids, while others are hard brittle solids depending on the relative prominence of their viscous and elastic characteristics. Many applications specifically exploit one or other of these but, in general, both contribute to the overall properties of the material and must be taken into account.

The mechanical properties of a polymer are investigated by loading a suitable specimen in some way and measuring the response. The interplay of time and temperature effects make the response of the material rather complicated. The experimental methods include both static (single motion) loading and dynamic (cyclic motion) testing, but the latter is often the best guide to the probable behaviour of a material under service conditions. For example, a plastic component in use is hardly likely to be stretched slowly until it nearly breaks but it is very likely to experience repeated small deformations. The greater the speed of testing the more brittle a specimen appears to be and the position of the T_g is naturally important. Perspex is brittle at ambient temperature but quite tough at elevated temperatures.

If a specimen is subjected to a constant stress (which does not produce fracture) there will normally be an immediate elastic response and this may be followed by a long slow creep in which the strain increases steadily perhaps to several 100 per cent. If, in a different kind of experiment, a specimen is rapidly strained to some predetermined length and maintained at that length, the necessary stress will be found to decay slowly, though probably never reaching zero. The interpretation of creep and stress relaxation, especially

FIG. 12. Idealized stress/strain curve for a ductile polymer, and appearance of test specimens at various stages. (Stress = extending force per unit cross-sectional area of specimen, strain = extension per unit of original length.)

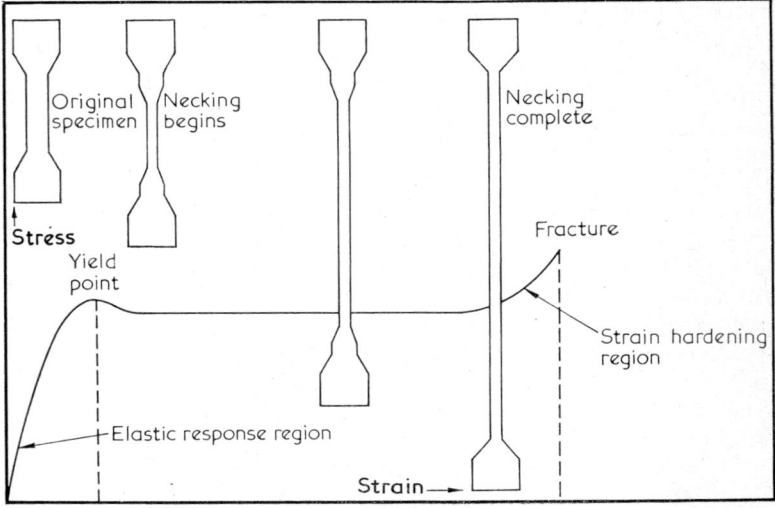

in terms of their temperature dependence, is a problem of great interest and much ingenious use has been made of theoretical models for viscoelastic behaviour.

It is even more difficult to account for the tensile stress/strain curve, an idealized version of which for a ductile polymer is shown in *Fig. 12.* In practice, the curve is obtained by stretching a bar-shaped specimen at a constant rate. The area under the curve is a measure of what is called toughness, and brittle polymers such as polystyrene only show the first rising part of the curve before fracture occurs. The curve in *Fig. 12* is common among semi-crystalline polymers such as nylon, and also among some amorphous polymers above their glass transition temperatures. Once the yield point is passed a considerable increase in strain occurs at virtually constant stress. The specimen 'necks' or narrows down, and the whole bar length gradually acquires the cross sectional area established where the neck began. In the final stage the stress must be increased once more to secure further increases in the strain because the material becomes stiffer just before fracture, an effect known as stress hardening. These features have all been at least qualitatively explained in molecular terms, and this inspires confidence in the molecular approach to major problems in polymer science. During necking some drastic alignment of the molecules occurs and strain hardening results from a locking together of aligned molecules.

5. Summary and Conclusions

This brief survey will have served its purpose if it has conveyed something of the nature of polymer science and the attractions of working in this area. For anyone interested in the subject as a basis for a career, a good foundation of chemistry, physics and mathematics is required, both at school level and later on. Within the broad subject of chemistry it is the study of kinetics, thermodynamics and molecular structure which will be most valuable to the young polymer scientist. Physics provides the foundation for an appreciation of mechanical properties and other matters, and the general applicability of mathematics is obvious. These basic subjects are so important that a case can be made for entering polymer science at post-graduate level, having first pursued a traditional major discipline for three years. Alternatively it is possible to take a first degree or other qualification in polymers at certain institutions and for some aspirants to careers in polymers this will be the appropriate route.

In addition to the topics which every properly qualified polymer scientist should know something about, there are many areas where other scientists with the right background can usefully apply their specialist knowledge. The understanding of Ziegler–Natta catalysis, for example, demands an expert grasp of organometallic stereochemistry, and the rheological description of molten polymers is the province of the applied mathematician firmly grounded in continuum mechanics.

For most people, a career in the polymer industry will involve technology as well as science. Polymer technology is a fascinating but difficult field where the principles of polymer science are harnessed to useful ends by the application of engineering principles. Success in the technological sphere presupposes a good general knowledge of the science on which it rests. For those who would like to broaden their knowledge of polymers before attending formal courses in the subject a number of general textbooks are available, and the Education Office of the Plastics Institute offers some interesting career literature.

In conclusion it may be said that the polymer industry offers an exceptionally wide range of activities, some of which are suitable for the well-educated polymer generalist and others for the specialist. At a time when the role of chemistry in society is under close scrutiny and when some feel that the subject has lost its way, a special word of encouragement to chemists is perhaps in order. The commercial synthesis of long chain molecules is one of the glories of modern chemistry, and the polymer industry as we know it today came into being as a result of the labours of some exceptionally imaginative chemists. Fittingly, many senior executives in the industry are chemists by training. If this tradition is to be maintained, chemistry must continue to prove that it is still the best discipline to provide a basis for the supply of high grade scientific nourishment on which the prosperity of the polymer industry will always depend.